ATTRACTING SCIENCE AND MATHEMATICS PH.D.S TO SECONDARY SCHOOL EDUCATION

Committee on Attracting Science and Mathematics Ph.D.s to Secondary School Teaching
Office of Scientific and Engineering Personnel Advisory Committee
Center for Education

AF834436

NATIONAL ACADEMY PRESS
Washington, DC

NOTICE: The project that is the subject of this report was approved by the Governing Board of the National Research Council, whose members are drawn from the councils of the National Academy of Sciences, the National Academy of Engineering, and the Institute of Medicine. The members of the committee responsible for the report were chosen for their special competencies and with regard for appropriate balance. Support for this project was provided by the Hewlett Foundation. Opinions, findings, conclusions, or recommendations expressed in this publication are those of the authors and do not necessarily reflect the views of the sponsors.

International Standard Book Number 0-309-07176-3
Additional copies of this report are available from: National Academy Press (http://www.nap.edu) 2101 Constitution Avenue, N.W., Box 285 Washington, D.C. 20055 800-624-6242 202-334-3313 (in the Washington metropolitan area)

Copyright 2000 by the National Academy of Sciences. All rights reserved.

Printed in the United States of America

THE NATIONAL ACADEMIES

National Academy of Sciences
National Academy of Engineering
Institute of Medicine
National Research Council

The **National Academy of Sciences** is a private, nonprofit, self-perpetuating society of distinguished scholars engaged in scientific and engineering research, dedicated to the furtherance of science and technology and to their use for the general welfare. Upon the authority of the charter granted to it by the Congress in 1863, the Academy has a mandate that requires it to advise the federal government on scientific and technical matters. Dr. Bruce M. Alberts is president of the National Academy of Sciences.

The **National Academy of Engineering** was established in 1964, under the charter of the National Academy of Sciences, as a parallel organization of outstanding engineers. It is autonomous in its administration and in the selection of its members, sharing with the National Academy of Sciences the responsibility for advising the federal government. The National Academy of Engineering also sponsors engineering programs aimed at meeting national needs, encourages education and research, and recognizes the superior achievements of engineers. Dr. William A. Wulf is president of the National Academy of Engineering.

The **Institute of Medicine** was established in 1970 by the National Academy of Sciences to secure the services of eminent members of appropriate professions in the examination of policy matters pertaining to the health of the public. The Institute acts under the responsibility given to the National Academy of Sciences by its congressional charter to be an adviser to the federal government and, upon its own initiative, to identify issues of medical care, research, and education. Dr. Kenneth I. Shine is president of the Institute of Medicine.

The **National Research Council** was organized by the National Academy of Sciences in 1916 to associate the broad community of science and technology with the Academy's purposes of furthering knowledge and advising the federal government. Functioning in accordance with general policies determined by the Academy, the Council has become the principal operating agency of both the National Academy of Sciences and the National Academy of Engineering in providing services to the government, the public, and the scientific and engineering communities. The Council is administered jointly by both Academies and the Institute of Medicine. Dr. Bruce M. Alberts and Dr. William A. Wulf are chairman and vice chairman, respectively, of the National Research Council.

www.national-academies.org

COMMITTEE ON ATTRACTING SCIENCE AND MATHEMATICS PH.D.S TO SECONDARY SCHOOL TEACHING

N. Ronald Morris, University of Medicine and Dentistry of New Jersey, *Chair*
Daniel Hamermesh, University of Texas
Kimberly Tanner, University of California, San Francisco
Kristina Peterson, Lakeside School, Seattle, Washington
Mary Long, University of Texas
Wendy Kopp, Teach for America

Staff

Marilyn Baker, *Associate Executive Director*
Peter Henderson, *Project Officer,* Office of Scientific and Engineering Personnel
George Reinhart, *Project Officer,* Office of Scientific and Engineering Personnel
Jan Tuomi, *Project Officer,* Center for Education
Sarah Choudhury, *Project Coordinator*
Cynthia Holmes, *Consultant*
Krystyna Isaacs, *Consultant*

OFFICE OF SCIENTIFIC AND ENGINEERING PERSONNEL ADVISORY COMITTEE

M. R. C. Greenwood (IOM), University of California, Santa Cruz, *Chair*
John D. Wiley, University of Wisconsin, *Vice-Chair*
Kenneth Arrow (NAS/IOM), Stanford University
Ronald Ehrenberg, Indiana University
Carlos Gutierrez, California State University, Los Angeles
Nancy B. Jackson, Sandia National Laboratories
Donald Johnson (NAE), Grain Processing Company
Martha A. Krebs, Institute for Defense Analysis
Stephen J. Lukasik, Independent Consultant
Claudia I. Mitchell-Kernan, University of California, Los Angeles
Michael Nettles, University of Michigan
Debra W. Stewart, Council of Graduate Schools
Tadataka Yamada (IOM), SmithKline Beecham Corporation
Thomas Young (NAE), Lockeed Martin Corporation (Retired)
Robert C. Richardson (NAS), Cornell University (*Ex Officio*)

Staff

Charlotte Kuh, *Executive Director*
Marilyn Baker, *Associate Executive Director*

CENTER FOR EDUCATION

Board of Overseers

Donald Kenndedy (NAS/IOM), Stanford University, *Chair*
Martha Darling, Independent Consultant, *Vice Chair*
Elizabeth Carvallas, Essex High School
Susan Fuhrman, University of Pennsylvania
Phillip Griffiths (NAS), Institute for Advanced Study
Dudley Herschbach (NAS), Harvard University
Richard Jager, University of North Carolina-Greensboro
Jo Anne Vasquez, Mesa Public Schools
Hyman Bass (NAS), University of Michigan (*Ex-Officio*)

Board of Advisors

Donald Kennedy (NAS/IOM), Stanford University, Chair
Martha Darling, Independent Consultant, Vice Chair
Jerome Grossman (IOM), Lion Gate Management Corp.
William J. Rutter (NAS), Chiron Corporation

PREFACE

For many reasons, the idea of attracting Ph.D.s to secondary school education suggests a difficult or, some would say, impossible concept. The need for secondary science and mathematics teachers is great, but is this the right group from which to recruit? Ph.D.s are not educated for this purpose, and most are probably better suited for careers in higher education, industry, or government that they have typically chosen. At the same time, they constitute a widely diverse population, with differing goals, personalities, and educational interests. And many share a desire to make science and math exciting to young people.

Our committee became involved in this study because we were interested in the potential benefits of brining more Ph.D.s into the secondary schools. For all the reasons stated in the report, we are now convinced that an effort to increase the number of Ph.D.s in the schools will make a significant qualitative contribution to the improvement of science and mathematics education. We also believe that Ph.D.s, who are trained to be inquisitive, to be creative, and to challenge established wisdom, will provide new leadership and be catalysts for change in science and mathematics education throughout their careers. We anticipate that the presence of Ph.D.s in the schools will provide an opportunity to generate new links between the K-12 system and the universites and will provide a new career path for Ph.D.s who seek to use their skills outside the traditional research setting.

This project has been organized in three phases. The current phase (phase one) is an exploration of the interest of Ph.D.s in secondary school positions and incentives that states, school districts, and others could use to attract them to such positions. Drawing on the results of phase one, and through a dialogue with those interested in secondary school science and mathematics education, the second phase will design demonstration programs to attract science and mathematics Ph.D.s to secondary school positions. Finally, the third phase will implement these demonstration programs to place Ph.D.s in such positions in selected states.

The first phase of this project was carried out under the auspices of the National Research Council's Office of Scientific and Engineering Personnel (OSEP), in conjunction with the Center for Education (CFE). To identify the potential interest of Ph.D.s in secondary school teaching and how they might be attracted to such positions, our committee carried out a national survey of graduate students and recent Ph.D.s. We also conducted a series of interviews with Ph.D.s working as secondary school teachers and educational administrators at the K-12 and postsecondary level. This report summarizes the findings from these investigations, with suggestions to the committee overseeing phase two of the project.

The charge of this committee was very narrowly directed to two issues: (1) is there a potential to attract science and mathematics Ph.D.s to secondary school teaching and other leadership positions, and (2) what incentives would be useful for states, school districts, and others in attracting them to such positions? The charge for the phase one committee was primarily to provide information to the committee overseeing the second phase of the project as it deliberates on how demonstration programs might be designed.

The committee was not asked to examine or substantiate the premises that underlie the charge, nor was it asked to implement its findings. Nor was it charged with assessing the potential benefits of placing science and mathematics Ph.D.s in secondary school teaching. These suggested benefits and their costs need to be made explicit and carefully evaluated by the phase-two study committee. However, the members of this committee have been convinced by our survey and interview data that a program to attract science and mathematics Ph.D.s to secondary school classrooms is potentially feasible, that it can be of significant value to the scientific and technical education of secondary school students, and that it will also provide a very rewarding new career opportunity for a select group of forward looking Ph.D. scientists and mathematicians.

N. Ronald Morris
Chair
Committee on Attracting Science and Mathematics
Ph.D.s to Secondary School Teaching

ACKNOWLEDGEMENTS

The *Committee on Attracting Science and Mathematics Ph.D.s to Secondary School Teaching* benefited from the contributions of many people. First and foremost, we wish to thank the Hewlett Foundation for providing funding for the initial phase of the project. We wish to thank all the graduate students and postdoctorates at Duke University, the University of Medicine and Dentistry of New Jersey, The University of Texas, the University of California, San Francisco, and the Fred Hutchinson Cancer Research Center who gave us their time and input by participating in focus groups. We also want to thank Michael Marder at the University of Texas, A. Leigh DeNeef at Duke University, and Nancy Hutchison at the Fred Hutchinson Cancer Research Center for their assistance in setting up the focus groups.

The project could not have been completed without the efforts of the two independent consultants who assisted with the project. They are Krystyna Isaacs, who conducted all interviews of Ph.D.s currently employed in science and mathematics secondary school education, and Cynthia Holmes, who facilitated all the focus groups and conducted all interviews of secondary school administrators. Finally, I wish to thank the members of the Committee for their contributions to the study and the final report.

This report has been reviewed in draft form by individuals chosen for their diverse perspectives and expertise in accordance with procedures approved by the National Research Council's (NRC) Report Review Committee. The purpose of this independent review is to provide candid and critical comments that will assist the study committee in making its report as sound as possible and to insure that the report meets institutional standards for objectivity, evidence, and responsiveness to the study charge. The review comments and draft manuscript remain confidential to protect the integrity of the deliberative process. I wish to thank the following for their participation in this review of the report: Camilla P. Benbow, Vanderbilt University; Roman Czujko, American Institute of Physics; Norma Davila, University of Puerto Rico; Sherry Hans, Pew Charitable Trust; Mozel Lang, Michigan Department of Education; Carolyn Morse University of North Carolina at Chapel Hill; Steve Rakow, University of Houston-Clear Lake; Gary Byrd, consulting engineer, report review moderator; and John Wiley, University of Wisconsin, Madison, report review coordinator. While the individuals listed above have provided constructive comments and suggestions, it must be emphasized that responsibility for the final content of this report rests entirely with the authoring committee and the institution.

The project was aided by the help of the NRC staff Marilyn Baker, OSEP deputy executive director; Peter Henderson, George Reinhart, and Jan Tuomi, project officers; Sarah Choudhury, project coordinator; and Rita Johnson, who provided editorial assistance.

N. Ronald Morris
Chair
Committee on Attracting Science and Mathematics
Ph.D.s to Secondary School Teaching

Executive Summary

The United States is at a critical juncture in science and mathematics education. The U.S. Department of Education has projected that the nation's school systems will need to hire more than two million new teachers during the next decade. Finding qualified teachers of science and mathematics will pose a special challenge, as many school districts already find it difficult to recruit science teachers. This report examines whether recent Ph.D.s in science and mathematics might provide an additional resource for helping to meet the nation's need for qualified secondary school science and mathematics teachers in the coming years, while creating valuable connections between U.S. schools and our vibrant science and engineering communities.

BACKGROUND

The National Research Council (NRC) has been deeply involved in the last decade in efforts to improve the science and mathematics education of our nation's schoolchildren. The 1996 *National Science Education Standards* urged "changes in what students are taught, in how their performance is assessed, in how teachers are educated and keep pace, and in the relationship between schools and the rest of the community—including the nation's scientists and engineers." It also emphasized the importance of "a new way of teaching and learning about science that reflects how science is done, emphasizing inquiry as a way of achieving knowledge and understanding about the world."

The NRC has followed the publication of the *Standards* with additional studies and programs that further explore key aspects of their implementation. In early 1999 the NRC launched a three-phase project to explore the feasibility of attracting scientifically trained Ph.D.s to positions in secondary school education as a possible mechanism to help improve science and mathematics education.

In launching this project, the NRC assumed that Ph.D. training, with its strong emphasis on experimental evaluation, quantitative approaches and mathematical content, could potentially make a meaningful contribution to the implementation of an inquiry-based learning environment. Also, due to the large number of Ph.D.s who have experienced difficulty recently moving out of postdoctoral positions—especially in the life sciences—the nation has an unusual opportunity to attract these Ph.D.s to America's secondary school classrooms. Most Ph.D.s are well suited to the research careers they have chosen and should continue to pursue them. Yet there are many Ph.D.s whose training, personalities, and outlook would make them ideal candidates for secondary school teaching positions.

There are, of course, a number of potential obstacles to Ph.D.s taking secondary school teaching positions. These include the willingness of Ph.D.s to take education courses and obtain certification; the attitudes of professors, colleagues, mentors, high school principals, and other secondary school teachers; the potential pposition of teachers' unions; salary levels; and others. The purpose of the first phase of the project was to evaluate the possibilities and obstacles and to recommend possible incentives to overcome them.

SCOPE OF STUDY

The NRC project to attract science and mathematics Ph.D.s to secondary education was organized into three phases. This first phase was designed to tell us whether there was any interest among Ph.D.s in becoming secondary school teachers and, if so, what incentives states and school districts might use to attract them. This report summarizes the findings from these investigations, with suggestions to the committee overseeing phase two.

Phase two will use information gained from phase one and other information sources to help design state-based demonstration programs to attract science and mathematics Ph.D.s to positions in K-12 education. Phase three will implement demonstration programs to place Ph.D.s in classrooms and possibly other educational positions in several test states. Phase three will also include an evaluation component to determine the effectiveness of the recruitment effort, as well as the potential benefits to the students taught by Ph.D. teachers.

The charge to the phase one committee was very narrowly directed. We were asked to determine (1) the likelihood that science and mathematics Ph.D.s could be attracted to secondary education, and (2) what special incentives might be useful for states, school districts, and others to attract Ph.D.s to such positions. The committee was not asked to examine or to substantiate the premises that underlie this project, nor was it charged with implementing its findings. Moreover, it was not charged with assessing the potential benefits of placing science and mathematics Ph.D.s in secondary school teaching. These suggested benefits and their costs should be made explicit and carefully evaluated by the phase-two study committee. The charge for the phase one committee was primarily to provide information to the committee overseeing the second phase of the project as it deliberates on how demonstration programs might be designed.

METHODS OF STUDY

To meet its charge, the committee investigated the career ambitions of Ph.D.s in the physical sciences, life sciences, and mathematics, and their willingness to take positions in secondary science and mathematics education under a variety of hypothetical conditions. Through focus groups and a national survey of graduate students and postdoctoral fellows, the committee investigated how teacher preparation programs, work conditions, and compensation packages could be modified to attract Ph.D.s to secondary school science and mathematics education.

In addition, the committee interviewed high school and magnet school principals, school district superintendents, state education policy makers, and graduate school deans to identify obstacles in the way of Ph.D.s taking secondary school positions as well as programmatic changes that could be used to attract Ph.D.s to secondary science and mathematics education. The committee also conducted interviews with Ph.D.s already working in secondary education to understand any barriers they had to overcome in taking these positions and their experiences in the secondary school environment. Committee members included experts in the life sciences and physical sciences, labor economics, graduate education, secondary school science teaching, teacher preparation, and partnerships between higher education and K-12 education institutions.

FINDINGS

Interest in Secondary Teaching

Based on our survey results, a large enough number of Ph.D.s appear to be sufficiently interested in secondary education for the NRC to explore a program to attract Ph.D.s to secondary school teaching positions. However, we cannot estimate the exact percentage of Ph.D.s who might ultimately consider secondary school teaching as a career. As with any career choice, this would depend on the specific incentives offered and the alternatives available at the time of choice. Our survey results demonstrate that the potential interest in careers in secondary school science and mathematics education is much higher than the 0.8 percent of Ph.D.s who currently work in K-12 education. The interest is high enough, we believe, to justify the development of demonstration programs to test the feasibility of this career alternative. This group of highly trained and knowledgeable individuals is potentially a valuable resource for secondary school science and mathematics education.

Respondents to our survey have typically considered at least four different options in contemplating their career futures; however, at least 36 percent of respondents have considered secondary school teaching or other secondary education positions in their career decision-making. Respondents who were still in graduate school, female, or U.S.

citizens were the most open to considering a career in secondary education. Chemists, with strong career options in industry, were less likely than respondents in the biological sciences, physics, and mathematics to consider secondary teaching positions.

Challenges and Possibilities

Given our survey results, a key question that a project designed to attract Ph.D.s to secondary school teaching must address is why, if up to 36 percent of science and mathematics Ph.D.s have considered secondary teaching careers, are less than 1 percent currently employed by K-12 educational institutions? We found that Ph.D.s have many negative perceptions about secondary school education that mitigate against their considering secondary school teaching positions. They perceive a lack of status and respect as teachers, poor classroom laboratory facilities, too many students in classrooms, structured curricula with little opportunity for creativity, possible conflicts with non-Ph.D. teachers, and student discipline problems. They also often perceive little value in education courses and see teacher certification as a barrier that is difficult to overcome. Low salary expectations for teaching in comparison to other careers also present a significant disincentive.

Stereotypes about Ph.D.s both in the secondary schools and in the universities create obstacles. For example, many school administrators argue that Ph.D.s may have good content knowledge, but do not have necessary pedagogical skills or cannot relate to secondary school students. In addition many university faculty do not promote nonacademic careers for Ph.D.s, much less careers in secondary school education. Indeed, graduate students typically aspire to positions in academic science and mathematics similar to those of their mentors, and the socialization process in graduate school strongly reinforces this career path.

Given these challenges, what do we, the phase one committee, believe is necessary for success? First, a program to attract Ph.D.s to secondary school teaching must combat negative perceptions about the secondary school environment. A first step would be to recruit Ph.D.s for whom the perceived positives outweigh the negatives. This may not be so difficult. While many participants in our focus groups held negative perceptions, many also held a number of positive perceptions, which included attractive working hours, a work schedule similar to their children's school day, and time for research or other activities during the summer. Many also believed they would enjoy the opportunity to foster the scientific interests of young people. Differences among focus group participants about whether they would prefer to teach at regular public high schools or science and technology magnet schools suggest that flexibility will be important if we want to attract Ph.D.s to teaching.

Our interviews with Ph.D. teachers and school administrators indicated that negative stereotypes about Ph.D.s as teachers are widespread, but have not posed obstacles to all Ph.D.s who have actually become teachers. In practice, only a minority of the Ph.D. teachers we spoke with had encountered resistance from school administrators or teachers. We also found that only a minority of the teachers we interviewed faced active disregard from colleagues and mentors after announcing a decision to take a secondary school position. We suggest that focusing on those states, school districts, schools, and graduate institutions where individuals—faculty and administrators—are most supportive of secondary school education as a potential career path for Ph.D.s would provide the most fertile ground for the demonstration projects.

We learned that those who select a secondary school teaching career should first assess their own personalities, interests, and skills. Those who will succeed and find fulfillment in secondary school education will be those who love teaching and enjoy helping students learn and achieve. Ph.D. teachers told us that for them the love of teaching and the enjoyment they get from working with children helps compensate for the higher salaries they could command in other science and mathematics careers. They also told us that preparation in educational pedagogy is essential, even for Ph.D.s, and that certification is an important outward sign of professional acceptance. Finally, it appears important that Ph.D.s contemplating teaching should have had some prior relevant experience and education to help them determine whether a teaching career is right for them.

Our data suggest that Ph.D.s can be attracted to secondary school education through programs that address their needs and interests and that help sustain them as teachers. Our survey presented graduate students and recent Ph.D.s with a number of scenarios, under which they were asked if they would consider secondary school science or mathematics teaching. Respondents indicated that they would be attracted by a fellowship program that provided training, placement, opportunities for networking with peers and that would cover their living expenses during the training period. They were strongly disinclined to undergo a normal full certification for teaching, but were quite amenable to an accelerated program. They were also interested in receiving mentoring during their classroom training. The potential availability of better resources for science education in the classroom and of better salaries were also of considerable interest to our survey respondents.

Other incentives that would serve to attract Ph.D.s to teaching careers would be access to a regional- or university-based science teaching resource center that provided science kits, loaned laboratory equipment, and organized field opportunities for science experiments in which students could participate. Many Ph.D.s would like to continue their involvement in science in some way. Our survey respondents who said they would consider secondary school careers indicated that having funding for summer research opportunities and for attendance at professional meetings during the school year is very appealing to them.

Although many survey respondents say they would be unwilling to make more than an initial two-year commitment to secondary school education, this should not be seen as limiting any activity designed to attract Ph.D.s to such positions. We believe it important first to recruit Ph.D.s to secondary education and then to work to retain them, as any industry tries to retain its valued employees. We anticipate that some of these individuals will discover their love for teaching children and remain in the program far beyond the initial two-year commitment. Among incentives for retention might be special training opportunities that would facilitate career options in K-12 leadership roles.

GUIDANCE FOR DEMONSTRATION PROGRAMS

As a next step, we recommend that the NRC continue to explore the development of demonstration programs. We recommend that the committee overseeing the second phase of the NRC's project on attracting Ph.D.s to secondary science and mathematics education convene a workshop consisting of interested persons in secondary school and postsecondary science and mathematics education to discuss in detail how demonstration programs might be structured. Based on the findings of the phase one study, we urge the committee overseeing phase two to consider the following programmatic features for demonstration programs described below.

ORGANIZATION OF DEMONSTRATION PROGRAMS

State Demonstration Programs. The committee overseeing the second phase of this project should consider developing demonstration programs in cooperation with a small number of interested states. State governments should organize these demonstration programs, because states play a stronger role than the federal government in education in the United States and can potentially bring more resources to bear than can local school districts. States would also develop their demonstration programs to fit their own educational and human resources needs.

Selection and Placement of Ph.D.s for Teaching Positions. The committee recognized that particular care must be taken in selecting Ph.D.s to participate in these demonstration programs. The selection process should identify individuals who have strong knowledge of their subject matter, a demonstrated interest in secondary school science and mathematics education, and personal characteristics appropriate to the secondary school environment.

Drawing from survey results, the committee suggests that state demonstration programs place and support Ph.D.s in a variety of secondary school education positions, including teaching positions in regular public secondary schools and science and technology magnet schools, as appropriate to the needs of the state. States should consider regional clustering of Ph.D.s in their demonstrations programs to facilitate networking, to optimize use of laboratory and science teaching resources, and to forge links between demonstration programs and university education and science departments.

Role of Postsecondary Institutions. The phase two committee should consider designing demonstration programs that have strong linkages to science and mathematics programs at colleges and universities in their states, piggybacking on any existing partnership programs. Colleges and universities could facilitate the recruitment and preparation of Ph.D.s for secondary school teaching and provide opportunities for classroom and secondary school experience for Ph.D.s interested in applying to the demonstration programs. They could also serve as venues for special workshops and meetings for Ph.D. teachers during the school year as part of a demonstration program in a given state. Finally, they could provide resources to support secondary school science and mathematics education.

Evaluation. Finally, we suggest that any demonstration program designed by the phase two committee include an evaluation component, to be implemented simultaneously with the demonstrations. The evaluation plan should address the feasibility of placing science and mathematics Ph.D.s in secondary school teaching, accessing the process of implementing such a program, and conducting an outcome evaluation based on measurable goals. A cross-site evaluation of the state demonstration programs, including their means for recruiting, placing, and supporting Ph.D.s in secondary school teaching, would inform other states considering similar programs.

PEDAGOGICAL SKILLS AND TEACHING RESOURCES

Education Courses and Certification. We strongly support the development of education courses and a teacher certification process tailored to the experiences and needs of Ph.D. scientists and mathematicians. Interviews with administrators and Ph.D. teachers indicated that education courses provide teachers with important pedagogical knowledge and that certification is an important step in establishing oneself as a teacher. We strongly agree. As a practical matter, however, courses leading to certification are not likely to be attractive to this population unless they can be accomplished in a fairly compressed manner. We found that forty-four percent of our survey respondents and more than two-thirds of those who had previously considered teaching careers indicated that they would consider teaching positions if they could receive their main training prior to beginning teaching by taking an intensive summer course in education. The percentage who would consider teaching if the period of time were increased to one year dropped precipitously to just 14 percent overall and 22 percent for those who had previously considered teaching. The Ph.D. teachers we interviewed indicated that they believed a streamlined course in educational theory and practice leading to certification could be developed for Ph.D.s.

We suggest that the state demonstration programs being designed by the phase two committee provide Ph.D.s with an intensive summer program in educational theory and practice as part of a process by which Ph.D.s could obtain teaching certification in an accelerated manner and should fund participants during this summer program. The summer program should focus on educational psychology, pedagogy, and pedagogical content knowledge.

Mentoring and Other Resources. Our survey clearly indicated that Ph.D.s would be more likely to consider teaching positions if they were mentored. The committee suggests that states should consider the selection and appointment of experienced teachers to serve as mentors to Ph.D.s participating in the demonstration programs. Providing mentors may add programmatic costs if states provide additional compensation to mentors, but the availability of mentors could be an important part of any program to introduce Ph.D.s to teaching careers.

We also found that 52 percent of survey respondents and three-quarters of respondents who had previously considered teaching would consider a secondary school teaching career if they received support from a regional- or university- based science resource center that provided science kits and loaned equipment or a partnership with an university. The committee overseeing the second phase of this project should work with states to determine whether the development of such science teaching resource centers would be a feasible component of state demonstration programs.

Future Positions for Ph.D.s. Ph.D.s could eventually contribute not only as teachers in the classroom, but also as leaders in other K-12 science and mathematics education positions. There was considerable interest among our survey respondents in providing professional development (e.g. teaching science or mathematics teachers), in becoming a science or mathematics specialist for a school district, in working in a university- or industry-based science educational partnership, or in serving as a science specialist in a science resource center. To a lesser degree, there was also interest in curriculum development or work in a science museum, environmental science center, or similar institution. While Ph.D.s could eventually contribute as leaders in K-12 science and mathematics education through these positions, we believe that it is essential for them to have secondary school teaching experience first.

INCENTIVES

The survey identified a number of incentives that respondents indicated would favorably affect their consideration of taking a position in secondary school teaching.

National Fellowship Program. Two-thirds of our survey respondents and almost 90 percent of respondents who had previously considered secondary school careers would consider taking a position as a secondary school teacher if they were awarded a fellowship that provided training, placement, and special opportunities for networking with peers, and covered living expenses during the training period. Given the potential such a fellowship might have for attracting Ph.D.s to secondary school teaching, the phase two committee should consider ways in which a fellowship program might be established and administered by a prestigious national agency or organization. The national program, instituted in cooperation with the states, could select and train fellows, fund them during their training, and provide an on-going opportunity for networking with peers.

A program that is national in scope could potentially attract resources from national sources, draw applicants from across the country, and serve as a catalyst for the state demonstration programs. A prestigious fellowship program would attract applicants who might not otherwise consider secondary school positions and produce a cohort of science and mathematics teachers who could conceivably change the way science and mathematics are taught. The potential downside to a "prestigious" fellowship for Ph.D.

teachers is that it might adversely differentiate them from the population of teachers we want them to join. We also recognize that the establishment of a national fellowship program would increase costs.

Compensation. Survey respondents recognized that salaries for secondary school teaching were lower than for other career options. Still, the average starting salary for teachers anticipated by graduate students and Ph.D.s in our survey—$37,400—is within the range of starting salaries offered to Ph.D. teachers by school districts, albeit at the high end of the range. For these programs to succeed, we believe that states and school districts will need to demonstrate a strong financial commitment by supplementing Ph.D. salaries. This might be done by providing stipends for attendance at scientific meetings and for other activities related to the professional development of the Ph.D.s and the benefit of their students.

We asked our survey respondents if they would consider a secondary school teaching position if they were guaranteed a two-year postdoctoral research fellowship at the end of a two-year teaching position, or if one year of their student loans were forgiven for each year of employment in a full-time teaching position. Given the low favorable response to these scenarios and the additional cost burden that they would place on a national program, we do not recommend that such incentives be offered as part of a national fellows program.

Peer Networking. Survey respondents indicated that, as teachers, they would welcome the opportunity to continue to network with their professional peers. Those surveyed responded very positively to consideration of teaching if they were provided opportunities for networking with peers. A fellowship program could include, among other devices, an annual meeting of fellows; participating states should also consider a regional clustering of Ph.D.s to facilitate networking opportunities.

Connections with the Larger Scientific Community. In designing state demonstration programs the phase two committee should consider providing opportunities for interactions between Ph.D. teachers and the scientific community in academia and industry. Respondents were very likely to consider teaching if they were given funding and time to attend at least one scientific meeting during the school year. We also found that respondents would consider teaching if they were guaranteed a summer fellowship, with travel expenses, in a research laboratory. The phase two committee should consider how states could develop links to universities and businesses to provide summer research opportunities for Ph.D.s, as they already do for science teachers in some states. Finally, the scientific community will need to provide these Ph.D.s with support and treat them as colleagues throughout their careers.

In a world filled with the products of scientific inquiry, scientific literacy has become a necessity for everyone. Everyone needs to use scientific information to make choices that arise every day. Everyone needs to be able to engage intelligently in public discourse and debate about important issues that involve science and technology. And everyone deserves to share in the excitement and personal fulfillment that can come from understanding and learning about the natural world. Scientific literacy also is of increasing importance in the workplace. More and more jobs demand advanced skills, requiring that people be able to learn, reason, think creatively, make decisions, and solve problems. An understanding of science and the processes of science contributes in an essential way to these skills. Other countries are investing heavily to create scientifically and technically literate work forces. To keep pace in global markets, the United States needs to have an equally capable citizenry.

-- National Science Education Standards, 1996

1

Introduction

The United States is at a critical juncture in its efforts to improve science and mathematics education for its children. Reports from the U.S. Department of Education suggest that the nation will need more than two million new teachers during the next decade. Finding qualified teachers of science and mathematics within this pool poses a special challenge. The nation needs to meet these challenges at a time when the National Research Council (NRC) and others are urging states and school districts to improve science and mathematics education through the implementation of *National Science Education Standards* and other efforts (National Research Council, 1996; National Research Council, 1997). This report examines whether recent Ph.D.s in science and mathematics might provide a resource for helping to meet the nation's need for qualified secondary school science and mathematics teachers in the future.

BACKGROUND

The Need for Qualified Science and Mathematics Teachers

We estimate that the nation will need to hire about 20,000 public secondary school science and mathematics teachers each year over the next decade, or about 200,000 in all. Those public secondary school teachers who spent the largest portion of their time teaching science and mathematics account for about 10 percent of all K-12 teachers (National Education Association, 1997). As shown in Table 1-1, the National Center for Education Statistics (NCES) has projected that the nation will need to hire more than two million new teachers from the 1998-99 school year to the 2008-09 school year. This projection is based on total public and private elementary and secondary school enrollment increasing 4 percent, from 52.2 million in 1997 to 54.5 million in 2006, and a large cohort of teachers retiring in the coming decade. (National Center for Education Statistics, 1999a; National Center for Education Statistics, 1999b). If one assumes that 10 percent of new hires will also be in secondary school science and mathematics, then we would expect to hire about 20,000 such teachers per year.

The need to increase the pool of well-qualified teachers in science and mathematics can be further seen in several places. For example, even while the need for new hires will grow over the next decade, finding qualified teachers of science and mathematics is already difficult for many school districts and may become an even greater challenge in the years ahead. Current data about underqualified teachers are disturbing; Richard Ingersoll reports that 26.6 percent of secondary (grades 7-12) mathematics students and 16.5 percent of secondary science students in the United States are taught by teachers without a major or minor in the field they are teaching. These percentages are much higher in urban and smaller rural school systems. (Ingersoll, 1999).

TABLE 1-1 Estimated Number of Newly Hired Full-time Equivalent Public School Teachers Needed Under Alternative Scenarios, 1994-1995 to 2008-2009 (in thousands)

	Scenario 1	Scenario 2	Scenario 3
	Constant pupil/teacher ratio[a]	Constant number of teachers[b]	Projections of Education Statistics to 2008[c]
Total needed 1998-99 to 2008-09 (11 years)	2,399	2,159	2,693
1994-95[d]	220	220	220
1995-96[d]	220	220	220
1996-97	223	177	224
1997-98	220	177	233
1998-99	214	180	218
1999-2000	210	181	227
2000-2001	212	184	235
2001-02	214	188	233
2002-03	218	192	244
2003-04	217	196	252
2004-05	221	199	253
2005-06	224	203	256
2006-07	224	208	256
2007-08	225	212	259
2008-09	221	215	261

NOTE: These estimates are based on the continuation rate of teachers from the 1993-94 school year to the 1994-95 school year.

[a] Number of teachers for 1996-97 through 2008-09 were produced by dividing the public school enrollment projections from the *Projections of Education Statistics to 2008* by the 1995-96 pupil/teacher ratio.

[b] Total number of teachers for 1996-97 through 2008-09 based on maintaining 1995-96 level.

[c] Total number of teachers for 1996-97 through 2008-09 is from the *Projections of Education Statistics to 2008,* which assumes decreasing class size.

[d] The number of newly hired teachers was computed using the actual number of teachers.

SOURCE: *National Center for Education Statistics 1999, Table 7, p. 35.*

The Third International Mathematics and Science Study (TIMSS) in the mid-1990s found, in assessing the performance of a half-million students around the world, that students in the United States were not performing at a level similar to their peers in other countries. A recent NRC report examining TIMSS data concluded, "the results of TIMSS suggest that U.S. students are falling short. Although U.S. fourth graders compare favorably to their international peers, U.S. eighth graders and high school seniors achieve at a lower level than do students in many other countries." The report argues that many factors affect the scores of U.S. students in TIMSS, including curricular issues, instructional practices, and school support systems. With regard to the latter, the report argued that one of several "particularly important aspects of this broader culture" is the preparation and support of teachers (National Research Council, 1999).

Similarly, the attrition rate for students who initially major in science, mathematics, and engineering as undergraduates is unacceptably high in 4-year institutions in the United States. Elaine Seymour and Nancy Hewitt recently found that more than 40 percent of males and 50 percent of females who initially majored in science, mathematics, and engineering fields switched to other majors, and that the problem was even more severe for minorities. For 40 percent of these field switchers, the authors found that inadequate high school preparation in subject matter and study skills was an important factor (and Hewitt, 1997). The quality of teaching and teaching methods play a central role in ensuring adequate subject matter preparation in high school.

Finally, the NRC's *National Science Education Standards,* published in 1996, urged dramatic changes in schools and school systems, including teacher preparation. This report advocates changes in how teachers are educated and keep pace and specifically recommends that teachers focus on "a new way of teaching and learning about science that reflects how science is done, emphasizing inquiry as a way of achieving knowledge and understanding about the world." (National Research Council, 1996). The need for more qualified secondary school science and mathematics teachers and the call for inquiry-based teaching underlie the NRC efforts to attract Ph.D.s to secondary school science and mathematics education.

An Untapped Resource

At the same time that the nation requires more qualified teachers of science and mathematics, the number of new science and mathematics Ph.D.s who have difficulty securing positions in academe or industry is growing. As shown in Table 1-2, there were more than 30,000 postdoctoral fellows in science and mathematics—more than 20,000 postdoctoral fellows in the life sciences alone—in the United States in 1998. Many of these postdoctoral fellows find themselves in a "holding pattern" of three or more consecutive postdoctoral fellowships while awaiting employment (National Research Council, 1998).

TABLE 1-2 Number of Postdoctorate Fellows, by Field, Physical and Life Sciences, 1998

Field	Total	M.D.s[a]	Ph.D.s[b]
Physical and life sciences, total	34,702	6,672	28,030
Physical sciences	6,869	26	6,843
Physics and astronomy	2,256	20	2,236
Chemistry	3,716	4	3,712
Earth, atmospheric, and ocean sciences	897	2	895
Mathematical sciences	274	16	258
Life sciences	27,559	6,630	20,929
Biological sciences	15,480	1,729	13,751
Health sciences	12,079	4,901	7,178

[a]Includes M.D., D.O. D.D.S., D.V.M. Also includes M.D.-Ph.D.s
[b]Does not include M.D.-Ph.D.s

SOURCE: *National Science Foundation/SRS, Survey of Graduate Students and Postdoctorates in Science and Engineering*

Recent data show that the job market may be improving as increasing numbers of new Ph.D.s—even in the life sciences—are entering tenure-track faculty positions at the time they receive their degrees. However, the job market for Ph.D.s and our system of graduate education operate such that supply and demand are out of synch from time-to-time, resulting in a periodic oversupply of science and mathematics Ph.D.s. Unfortunately, those already stuck in postdoctoral positions—especially those who have been in more than one—do not appear to be benefiting from the improving job market. Thus, the large numbers of Ph.D.s who have found themselves in this postdoctoral "holding pattern"—the current "oversupply" of Ph.D.s—may provide an untapped source of qualified secondary school science and mathematics teachers. This study aims to explore the potential interest of Ph.D.s in these positions.

NRC ROLE IN SCIENCE AND MATHEMATICS EDUCATION

The National Research Council has been deeply involved in the last decade in efforts to improve the science and mathematics education of our nation's schoolchildren, emphasizing achieving the national goal of scientific and mathematical literacy for all of our students. The *National Science Education Standards,* published in 1996, urged dramatic changes in schools and school systems. The *Standards* advocate "changes in what students are taught, in how their performance is assessed, in how teachers are educated and keep pace, and in the relationship between school and the rest of the community—including the nation's scientists and engineers." They also emphasize "a new way of teaching and learning about science that reflects how science is done, emphasizing inquiry as a way of achieving knowledge and understanding about the world."

The NRC has since continued to initiate studies and programs designed to make a difference in our nation's science and mathematics education at all levels. The Center for Education (CFE) leads this effort by examining the efficacy and influence of standards-based systemic reform, the preparation and professional development of teachers, and programmatic and curricular models for improved mathematics, science, and engineering education. The NRC is also exploring the science of learning, the skills of teaching, and the assessment of student achievement. To this end the Center produces reports on implementing science standards, teacher preparation, and student assessment designed, above all, to improve science and mathematics education. Recent NRC reports include:

Every Child a Scientist: Achieving Scientific Literacy for All (1997)

Improving Student Learning in Mathematics and Science: The Role of National Standards in State Policy (1997)

Improving Teacher Preparation and Credentialing Consistent with the National Science Education Standards: Report of a Symposium (1997)

Science Teacher Preparation in an Era of Standards-Based Reform (1997)

Teaching About Evolution and the Nature of Science (1998)

Designing Mathematics or Science Curriculum Programs: A Guide for Using Mathematics and Science Education Standards (1999)

Global Perspectives for Local Action: Using TIMSS to Improve U.S. Mathematics and Science Education (1999)

How People Learn: Bridging Research and Practice (1999) Commission on Behavioral and Social Sciences and Education

Improving Student Learning: a Strategic Plan for Education Research and Its Utilization (1999) Commission on Behavioral and Social Sciences and Education

Inquiry and the National Science Education Standards: A Guide for Teaching and Learning (2000)

PROJECT ON ATTRACTING PH.D.S TO SECONDARY EDUCATION

In early 1999 the NRC launched a three-phase project to explore the feasibility of attracting science and mathematics Ph.D.s to positions in secondary school education. The essential goal of the project is to improve the science and mathematics education and literacy of our nation's schoolchildren. Among the keys to ensuring ongoing improvement in science and mathematics education—standards-based systemic reform, the excellent preparation and professional development of teachers, new programmatic and curricular models for improved science and mathematics education, and the appropriate assessment of student achievement—this project deals with the supply, preparation, and professional development of teachers. It does so by focusing on attracting those Ph.D.s who could be well suited to secondary school education positions and could bring new techniques—such as inquiry-based learning—and a strong knowledge of science and mathematics to the classroom. This study reports on the outcomes of the first phase of the project, which focuses on assessing the potential interest of Ph.D.s in such positions and the changes that states, school districts, or others might implement to attract these Ph.D.s.

Underlying Assumptions

In launching this project, the NRC made assumptions about the contributions that Ph.D.s could bring to secondary schools. These were: (1) that Ph.D. training with its strong emphasis on experimental evaluation of scientific questions could be useful to teachers in providing the inquiry-based learning environment in schools recommended in the *National Science Education Standards,* and (2) that the deeper training of Ph.D.s in quantitative approaches and content—and their firm attachments to the nation's scientists and engineers—could also be very helpful in improving American science and mathematics education.

The project also began with the assumption that the job market for Ph.D.s and our system of graduate education are such that supply and demand for researchers will be out of phase from time-to-time, resulting in a periodic oversupply of science and mathematics Ph.D.s who might be an untapped resource for science and mathematics teaching. In 1997, an NRC report on the careers of recent Ph.D.s in the life sciences indicated that the number of postdoctoral fellows in the United States has been growing and that—particularly in the life sciences—many recent Ph.D.s have experienced difficulty moving out of postdoctoral positions into permanent careers in either academia or industry. Thus, we have reached one of those moments when an oversupply of Ph.D.s has occurred and the nation has an unusual opportunity to attract these Ph.D.s to America's secondary school classrooms.

The committee established to oversee the first phase of this project believes that most Ph.D.s are well suited to the research careers they have chosen and should continue to pursue them. Yet there are many Ph.D.s whose training, personalities, and outlook would make them ideal candidates for secondary school teaching positions and potential leaders in science and mathematics education. Just as there are select groups of science and mathematics Ph.D.s who have gone on to other careers outside of postsecondary teaching and research, we anticipate that secondary education may also provide attractive career opportunities for some Ph.D. scientists. Indeed, since 1991, a majority of Ph.D.s in science and engineering have worked in careers outside of postsecondary academic institutions (National Academy of Sciences, 1995; Rice, 1996; Tobias et. al., 1995; National Science Foundation, 1996; Commission on Professionals in Science and Technology, 1997; Association of American Universities, 1998).

There are, of course, a number of potential obstacles to Ph.D.s taking secondary school teaching positions. These include the willingness of Ph.D.s to take education courses and obtain certification; the attitudes of professors, colleagues, mentors, high school principals, and other secondary school teachers; the potential opposition of teachers' unions; and salary levels. It is the object of the phase one study to explore the interest of Ph.D.s in secondary school education and to see what incentives might be implemented to overcome the potential obstacles to Ph.D.s following this career path.

Problems and Possibilities

The NRC launched this project understanding that certain aspects of the graduate education and secondary school environments pose obstacles to Ph.D.s taking secondary school teaching positions. Success will depend, first, on the willingness and ability of Ph.D.s to apply their knowledge and skills in a different educational environment. Success will also depend on their ability to obtain certification, possibly through alternative certification programs that have already provided teacher preparation to many others. School districts, states, and higher education institutions will need to design programs—including appropriate teacher preparation programs —that offer incentives for scientifically trained Ph.D.s to enter secondary school education. Finally, the scientific community will need to provide these Ph.D.s with support and treat them as colleagues throughout their careers.

We have some information about areas where we might meet obstacles to success.

Number of Ph.D.s Employed in K-12 Educational Institutions. As displayed in Table 1-3, data from the 1997 Survey of Doctorate Recipients (SDR) show that only a very small fraction—just 0.8 percent—of Ph.D.s in the life sciences, physical sciences, and mathematics currently work in K-12 educational institutions.

TABLE 1-3 Ph.D.s in Biological and Health Sciences, Physical Sciences, and Mathematics Employed in K-12 Educational Institutions, 1997

	Employed	Employed in K-12 Education*	
Field	Population	Estimated Percent	Estimated Number
Physical and life sciences, total	275,860	0.8%	2,200
Physical sciences	129,650	0.8%	990
Physics and astronomy	35,920	0.8%	280
Chemistry	54,220	0.9%	470
Earth/atmospheric/marine sciences	15,110	0.7%	110
Mathematical sciences	24,400	0.5%	130
Life sciences	121,810	0.9%	1,080
Biological sciences	104,630	0.9%	900
Health sciences	17,180	1.0%	180

*These data are calculated by special tabulations of a sample survey. They represent approximate estimates of the number and percentage of science and mathematics Ph.D.s employed in K-12 educational institutions.

SOURCE: *National Science Foundation/Survey of Doctorate Recipients, special tabulation.*

Alternative Certification. For nearly 20 years, alternative certification has helped school districts and states meet the need for qualified teachers and will continue to be important as they seek to hire 200,000 or more teachers each year in the next decade. School districts, states, and institutions of higher education first developed alternative certification programs in response to the threat of teacher shortages in the early 1980s and the number of programs has grown since. In 1998, according to the private, non-profit National Center for Education Information (NCEI), 41 states and the District of Columbia report at least one alternative teacher certification program. These programs include a wide range of efforts, as NCEI notes, "from emergency certification to very sophisticated and well-designed programs that address the professional preparation needs of the growing population of individuals who already have at least a baccalaureate degree and considerable life experience and want to become teachers" (Feistritzer and Chester, 1998).

Many individuals with non-traditional backgrounds have and could contribute their knowledge and skills to secondary school education. NCEI estimates that more than 80,000 individuals—those trained to teach years ago but who never did, those who switch careers or retire from the military, and those from other institutions such as liberal arts colleges—have already been certified through these programs. Thousands more receive certification through alternative programs created by institutions of higher education (Feistritzer and Chester, 1998).

Graduate School Training. Although many science graduate programs require that their Ph.D. candidates teach undergraduates as part of their training, none of these programs currently view its mission to prepare graduates to teach in secondary schools. Indeed, graduate mentors may explicitly or implicitly discourage their students from spending too much time and effort preparing for careers in teaching if it "distracts" them from their research projects or lengthens the time needed for them to obtain their degree (Kennedy, 1997; Boyer Commission on Educating Undergraduates in the Research University, 1998).

In spite of the obstacles, the potential advantages of attracting individuals with advanced scientific training into undergraduate and secondary education have not gone unnoticed. Two new programs seek to improve the preparation of graduate and postdoctoral students who wish to assume teaching positions at undergraduate institutions. These are the Preparing Future Faculty (PFF) program, funded by the Pew Charitable Trust, and Shaping the Preparation of Future Science and Mathematics Faculty, funded by the National Science Foundation. These programs, sponsored by the Association of American Colleges and Universities and the Council of Graduate Schools, encourage colleges and universities to provide graduate students with experience in developing and teaching undergraduate courses outside of their own institutions.

The National Science Foundation has recently recognized the potential for the pool of doctorate-level talent to work in pre-college education. In February 1999, the NSF announced the creation of the Graduate Teaching Fellows in K-12 Education (GK-12) program, which provides graduate students and advanced undergraduates in science the opportunity to receive training, gain experience, and serve as resources for science education in K-12 schools. Expected outcomes from this new initiative include "…improved communication and teaching skills for the fellows, enriched learning by K-12 students, professional development opportunities for K-12 teachers, and strengthened partnerships between institutions of higher education and local school districts" (National Science Foundation, 1999).

Organization of the Project

The project has been organized into three phases: (1) an exploration of the interest of Ph.D.s in secondary school positions and incentives that states, school districts, and others could use to attract them to such positions; (2) the design of programs to attract science and mathematics Ph.D.s to secondary school positions, drawing on the results of phase one, and a dialogue with those interested in secondary school science and mathematics education; and (3) the implementation of demonstration programs that place Ph.D.s in such positions in selected states.

Phase One: Assessing the Potential Interest of Ph.D.s in Secondary School Science and Mathematics Education

The first phase of the project was carried out under the auspices of the NRC's Office of Scientific and Engineering Personnel (OSEP), in conjunction with the Center for Education (CFE). To identify the potential interest of Ph.D.s in secondary school teaching and how they might be attracted to such positions, this committee carried out a national survey of graduate students and recent Ph.D.s. We also conducted a series of interviews with Ph.D.s working as secondary school teachers and educational administrators at the K-12 and postsecondary level. This report summarizes the findings from our investigations, with suggestions to the committee overseeing phase two.

Phase Two: Designing Demonstration Programs

In the second phase of this project, a new committee will bring information from a number of sources to bear on the design of programs to attract Ph.D.s to secondary school science and mathematics education. One source of information will be the data and recommendations contained in this report. Another will be a workshop of interested persons in secondary school science and mathematics education—including representatives of states interested in sponsoring demonstration programs—to discuss the design of demonstration programs. Based on the outcomes of that workshop, discussions with other experts, and its own deliberations, the phase two committee will produce a report that recommends components of demonstration programs.

Phase Three: Implementing State Demonstration Programs

Once phase two of the project has been completed, the NRC will seek funding to implement and evaluate demonstration programs in three to five states, based on the goals and design elements identified in phase two. This phase will also be overseen by an NRC committee.

SCOPE OF THE STUDY

The Charge to the Phase One Committee

The charge to this committee was very narrowly directed to two issues:

(1) Is there a potential to attract science and mathematics Ph.D.s to secondary school teaching and other leadership positions?
(2) What incentives would be useful for states, school districts, and others in attracting them to such positions?

The committee was not asked to examine or substantiate the premises that underlie the charge, nor was it asked to implement its findings. Moreover, it was not charged with assessing the potential benefits of placing science and mathematics Ph.D.s in secondary school teaching. These suggested benefits and their costs should be made explicit and carefully evaluated by the phase-two study committee. The charge for the phase one committee was primarily to provide information to the committee overseeing the second phase of the project as it deliberates on how demonstration programs might be designed. Nevertheless, members of this committee do believe that a program to attract science and mathematics Ph.D.s to secondary school classrooms could be of significant value to science education of secondary school students and would also provide a rewarding new career opportunity for Ph.D. scientists.

The charge to the committee was to investigate the career ambitions of Ph.D.s in the physical and life sciences through focus groups, a national survey, and interviews. The committee was also asked to examine the potential interest of recent Ph.D.s in secondary school teaching and curriculum development positions and to determine the kinds of incentives—regarding work conditions and compensation packages—that states, school districts, and others could use to induce Ph.D.s to take these positions. The positions included the teaching of physics, chemistry, biology, and various electives in public high schools, as well as positions fostering better science and mathematical education at the secondary level.

The committee found it necessary to discuss and address two issues that developed directly from the study data and that extended beyond its specific charge. First, in the course of holding focus group discussions, the committee found that many participants believed there were a number of ways, beyond secondary school teaching and curriculum development, that Ph.D.s could contribute to improving science and mathematics education. Graduate students and postdoctoral fellows clearly liked the idea of working on the development of science and mathematics curricula, not just for secondary school courses but across grades from kindergarten to grade 12. Focus group participants suggested that Ph.D.s could also contribute by working in professional development, science education partnerships, or science museums and environmental centers. They indicated that Ph.D.s might be interested in serving as science or mathematics specialists in school districts, science resource teachers in elementary or secondary schools, or science specialists in regional or university-based science education resource centers.

The committee believed it was important to explore these suggestions both in their own right and as context for the more specific questions on secondary school teaching and curriculum development it was asked to address. While the focus of this study remains secondary school science and mathematics education, the suggestions of focus group recipients were addressed by adding an additional question to the survey questionnaire. The results of this question are discussed at the end of Chapter 2.

Second, the committee realized that recent Ph.D.s by themselves cannot satisfy the need for new secondary school science and mathematics teachers. We considered the question of whether we should also examine the interest of other individuals with extensive science and mathematics education backgrounds. This included those who were ABD ("all but dissertation") and/or had master's degrees. In this case, the committee decided that a systematic exploration of the interest of these individuals in secondary school teaching and how to attract them to positions would take them beyond their charge.

Nonetheless, the committee believes that graduate students in science and mathematics who obtain master's degrees or who are "ABD" should also be considered a potential source of qualified science and mathematics teachers. The fact that the focus groups we held and the questionnaire we fielded during the course of this study included graduate students as participants and respondents in some measure substantiates the potential interest of these additional groups, as well as the interest of those who have already received the Ph.D, degree.

Questions to be Answered by the Present Study

The data reveals that there is only a small number of Ph.D.s working in K-12 educational institutions, but is the potential interest of Ph.D.s in working in secondary school education higher than this small number might suggest?

How do Ph.D.s perceive the teacher certification process and what is required to ensure that Ph.D.s become certified in an expeditious and effective manner?

How should Ph.D.s who take positions in secondary school science and mathematics education be prepared?

What incentives—related to the work environment and compensation—might be useful in attracting Ph.D.s to positions in secondary school science and mathematics education?

What have been the experiences of Ph.D.s who have taken secondary school teaching and curriculum development positions and what can we learn from these experiences?

How do secondary school and higher education administrators perceive the strengths and weaknesses of Ph.D.s as secondary school science and mathematics teachers?

What kinds of ongoing relationships should Ph.D.s who teach secondary school science and mathematics have with postsecondary education and academic research?

This study is a first effort at addressing these kinds of questions. The information gathered here is designed to inform the discussions and deliberations of the committee overseeing the second phase of this project as this committee addresses options for developing demonstration programs with a small number of states.

METHODS OF STUDY

The committee began its information gathering by conducting five focus groups with advanced graduate students and recent Ph.D.s on their career aspirations and perceptions of secondary school science and mathematics education careers. Led by a professional facilitator, the focus groups obtained reactions to a number of hypothetical incentives that might be used to attract Ph.D.s to secondary school positions. The groups provided background information and elicited topics for a survey questionnaire to be fielded to graduate students and recent Ph.D.s on their career aspirations.

To obtain data on the willingness of Ph.D.s to consider positions in secondary school science and mathematics education and on the various conditions that might enhance their recruitment, the committee conducted a national survey of graduate students and postdoctorates. The survey questionnaire reflects input from the five focus groups as well as the committee members and staff from several NRC units. It covers four broad areas of interest: (1) the demographic characteristics of the respondents, (2) their short-term and long-term career aspirations, including salary expectations, (3) the conditions under which the respondent might consider secondary school science and mathematics education as a career, and (4) incentives that might be required for the respondents to consider such careers. Respondents were asked to comment on their interest in a range of positions in K-12 science and mathematics education: elementary and secondary school teaching; K-12 science or mathematics curriculum development; professional development for teachers; school, district, or regional science resource specialist; working for a science education partnership; or working in a science museum, environmental center, or similar institution.

The survey was fielded in July and August 1999 to a national sample of 2,000 graduate students and recent Ph.D.s stratified by field of study. We received 719 responses, for a 39 percent response rate. A telephone survey of nonrespondents to the survey indicated that there was little or no detectable nonresponse bias. The respondents had demographic characteristics similar to the population of graduate students and recent Ph.D.s from which our sample was drawn. The number of responses provided sufficient power to perform the required data analyses.

In addition to fielding the survey, staff conducted a series of telephone interviews with interested individuals in secondary school science and mathematics education during the summer of 1999. We conducted interviews with 18 science and mathematics Ph.D.s currently in secondary school teaching or curriculum development positions to learn about the barriers they encountered, their strategies for overcoming these barriers, their areas of success and achievement, and their thoughts on the recruitment and retention of other science Ph.D.s. Interviews were also conducted with high school principals, school district superintendents, and chief state school officers to obtain information about their attitudes toward employment of Ph.D.s in secondary schools, about certification issues, and about funding of Ph.D.s in secondary school education. In addition, telephone interviews were conducted with graduate school deans to identify potential obstacles and ascertain the kinds of programmatic changes required in graduate education to prepare Ph.D.s for careers teaching in secondary schools. A more detailed explanation of the study methodology is provided in Appendix A.

The committee deliberated whether staff should interview other interested individuals beyond the principals, superintendents, and deans, such as union representatives, non-Ph.D. secondary school teachers and postsecondary faculty, but determined that limited time and resources for its part of the overall project precluded interviewing these individuals. The committee overseeing the second phase of the project is charged with engaging the many additional interested persons, including union representatives, in conversations about bringing Ph.D.s into classrooms.

OVERVIEW OF THE REPORT

Chapter 2 reviews the results from five focus group sessions and from the national survey of graduate students and postdoctorates fielded during the summer of 1999. The data explore the willingness of recent Ph.D.s to consider secondary school education positions and the kinds of programs and incentives necessary to attract Ph.D.s to these positions. Chapter 3 presents the results of our interviews with Ph.D.s, secondary school administrators, and graduate deans. These results further inform our analysis of the opportunities and obstacles Ph.D.s face in pursuing careers in secondary school science and mathematics education. Chapter 4 summarizes the committee's findings and provides suggestions for the kinds of programs and incentives the committee overseeing the second phase of this project should consider in developing options for demonstration programs.

2

Will Ph.D.s Consider Careers in Secondary School Science & Mathematics Education?

To collect data on the career aspirations of Ph.D.s and to explore their potential interest in secondary school science and mathematics education careers, the committee fielded a Survey of Graduate Students and Recent Ph.D.s. We received more than 700 responses to the survey and this chapter presents an analysis of these responses. Most important, the survey results demonstrate that potential interest in teaching at the secondary school level appears to be much higher among Ph.D.s than the 0.8% of current Ph.D.s who are employed at the K-12 level would tend to indicate. Below we provide a description of respondents' current status and field of science or mathematics, their demographic characteristics and career aspirations, and factors that could affect the interest of Ph.D.s in careers in secondary school teaching and education. The survey questionnaire is presented in full in Appendix C.

SURVEY METHODS AND RESPONSE

Survey Methodology and Response

The survey questionnaire was initially fielded to 2,000 graduate students and recent Ph.D.s during the summer of 1999. A first mailing was sent in July 1999 with a second, follow-up mailing about three weeks later in August 1999. Of the original 2,000 questionnaires mailed, the postmaster returned as undeliverable 169 forms. Another nine individuals responded that they were neither graduate students nor recent Ph.D.s and thus were out of the scope of the survey. The net number of individuals to whom the survey was fielded, then, was 1,822, and we received from them 719 responses. This resulted in a response rate of 36 percent of the original sample and 39 percent of the in-scope individuals contacted.

We also conducted a follow-up phone survey of 200 randomly selected nonrespondents. The purpose of this follow-up survey was to assess the reason for their failure to respond. Specifically, we were interested in determining if these persons did not respond to the survey because they were negatively biased towards careers in secondary education for science and mathematics Ph.D.s. As detailed in the appendix, we were able to obtain phone numbers for 174 of these individuals, whom we attempted to contact in late August. When contacted, 22 replied that they had already completed and returned the survey and six indicated they were neither doctoral graduate students nor recent Ph.D.s. Of the remaining 146 individuals, 94 (64 percent) fell into one of four categories: (1) they had not completed the survey because they were out of the office when it arrived, (2) a colleague indicated that the respondent was on vacation, (3) they did not respond to a voice mail message, or (4) they did not answer at all. Only three persons stated that they were not interested in participating in the survey and none of the persons we contacted in the follow-up survey indicated that they failed to respond because they were negatively predisposed towards careers in secondary education.

A number of factors contributed to the overall in-scope response rate of 39 percent, which is lower than the committee would have liked. First, the committee was asked to undertake this study in a short period of time in 1999 as a first step before the second phase of the project began in early 2000. This timing obliged us to field the questionnaire in July and August, when many graduate students and postdoctoral fellows are not available on campus. Second, many graduate students and postdoctoral fellows are in transit from one position or residence to another during the summer. The committee was dependent on university departments to provide accurate and up-to-date addresses for the respondents, which they do not always have.

While we can say (as we do below) that the demographic characteristics of respondents to our survey are similar to those of the population we sampled, we cannot, of course, say for certain how non-respondents would have replied to the questionnaire. It is possible that, among those individuals who actually received the questionnaire, those who had an interest in secondary school education or who were at least neutral responded in higher proportions than those who had no interest. Such a result calls into question whether the respondents serve as a representative sample of the broader population of Ph.D.s. The results of this survey, therefore, should not be interpreted as the definitive word on the degree of interest of new Ph.D.s in the sciences and mathematics nationally in secondary school education. We do not know what that number or percentage might be and, as with any career choice, it most likely depends on the specific incentives offered and alternatives available at the time of choice. However, the survey results do demonstrate that potential interest in careers in prebaccalaureate education, even at its lower bounds, appears to be much higher among Ph.D.s than is represented by the 0.8 percent of them who currently work in K-12 education. This interest is high enough, we believe, to justify pilot programs to test the feasibility of this career alternative.

Characteristics of Respondents

By design, over half (57 percent) of the graduate students and recent Ph.D.s in our original sample were in the biological sciences, with the rest in physics, chemistry, earth, atmospheric, and marine sciences, and mathematics. We designed the sample in this way so it would mirror the universe from which it was drawn: according to the Survey of Earned Doctorates (SED), which collects data on new Ph.D. recipients population, there were 7,111 new Ph.D.s in the life sciences (biological sciences, biochemistry, and health sciences) in 1997 and 5,685 new Ph.D.s in the physical sciences (physics and astronomy, chemistry, earth, atmospheric, and marine sciences, and mathematics) in that year.

In general, the distribution of fields among our respondents reflects the distribution within the sample and the population. Over half (55 percent) were in the biological sciences, thus mirroring the overall population. However, one group in particular was underrepresented among our respondents. Only one-sixth (16 percent) of the respondents were chemists, though they made up 23 percent of our sample. Respondents in physics, earth sciences, and mathematics were 11 percent, 9 percent, and 8 percent, respectively, of our total respondants.

Three-fourths (75 percent) of the respondents were graduate students and one-fifth (20 percent) were Ph.D.s in postdoctoral positions. The remaining respondents had received their Ph.D. degree but were not in a postdoctoral position. Many of these respondents were employed in research in science departments in medical schools.

This distribution, again, is similar to the current distribution across these categories within the universe of people being addressed. According to the Survey of Graduate Students and Postdoctorates in Science and Engineering, the population in the universe we sampled consisted in 1998 of 118,452 graduate students, or 77 percent of the population, and 34,702 postdoctorate fellows, or 23 percent of the population.

A plurality (45 percent) of respondents was age 25 to 29, followed by respondents age 30 to 34 (28 percent). Only 12 percent of respondents were younger than 24 and only 14 percent were age 35 or over. This distribution is within the expected range given recent results from the SED. In 1998, the average age of new Ph.D.s in the physical sciences was 30.7 years and in the life sciences 32.3 years.

About the same percentage of respondents to our survey were single (49 percent) as were married (43 percent). According to the SED, the percent of physical scientists married at the time that they received their degrees was 50.6 percent in 1998 and the percentage of life scientists who were married was 56.2 percent.

Native born U.S. citizens comprised 62 percent of respondents and naturalized citizens or permanent residents about 10 percent. The 176 foreign respondents on temporary visas represented 46 different countries of origin; most of these temporary residents came from the People's Republic of China (46), India (20), and the Republic of Korea (10). This breakout of survey respondents by citizenship status—72 percent U.S. citizens or permanent residents, 24 percent temporary visa holders, and 4 percent unknown is nearly identical to the distribution of new Ph.D.s in the 1998 SED across these citizenship groups.

Nearly 70 percent of respondents classified themselves as white and about one-fourth as Asian/Pacific Islander. The number classifying themselves as African American and Native American was 18 (3 percent) and 17 (3 percent), respectively. Only 14 respondents indicated they were Hispanic. In 1998, about 53 percent of the all respondents to the SED, but 70 percent of U.S. citizens and permanent residents, classified themselves as white.

Finally, three-fifths (62 percent) of the respondents were male. In 1998, 62 percent of new Ph.D. recipients in the fields we surveyed were male according to the SED.

Based on these comparisons, we believe the respondents to our survey are very similar in demographic characteristics to the universe of graduate students and postdoctorates in the fields we are interested in. Only one group in our sample is noticeably under-represented among our respondents, and that is graduate students and postdoctorates in chemistry. These individuals comprised 23 percent of our sample, but only 16 percent of our respondents. As discussed below, chemists who responded to the survey were less interested in secondary school teaching and curriculum development positions compared to respondents in other groups.

CAREER ASPIRATIONS

For nearly half of all respondents, their next step in employment was a postdoctoral position. Nearly three-fifths (58 percent) of graduate students anticipated a postdoctoral position as their immediate next step in employment or training, while only 21 percent of postdoctoral fellows sought an additional postdoctoral position for their immediate next step. In addition, over one-fifth of respondents indicated employment in industry as their next step. A position in K-12 education was the next step for only 8 respondents. The distribution of immediate next steps in employment or training is shown in Table 2-1.

When asked about their ultimate career aspirations, over 60 percent of respondents selected careers in academia. The most popular career choice was a faculty position in a research university. A position in industry was the second most selected career aspiration, selected by over 24 percent of respondents. A position in K-12 education was the ultimate career aspiration for only 9 respondents. The distribution of respondent's ultimate career aspirations is shown in Table 2-2.

TABLE 2-1 Percentage Distribution of Immediate Next Step in Employment or Training by Respondent Status

Immediate Next Step In Employment or Training	Status of Respondents			
	Graduate Students (N=513)	Postdocs (N=134)	Other Ph.D.s (N=33)	All Respondents (N=680)
Postdoctoral position	58%	21%	13%	49%
Research university faculty position	3	36	29	11
Liberal arts college position	4	6	0	4
Community college faculty	1	0	3	1
Other academic position	3	8	3	4
K-12 education position	1	0	3	1
Non-profit organization	1	0	3	1
Government	3	2	0	2
Industry	20	24	39	22
Self employed	1	1	3	1
Further education	3	0	3	2
Other position	2	3	1	3
Total	100%	100%	100%	100%

TABLE 2-2 Percentage Distribution of Respondent's Ultimate Career Aspirations

Ultimate Career Aspiration	Percent of All Respondents (N=680)
Postdoctoral position	1%
Research university faculty position	43
Liberal arts college position	14
Community college faculty	1
Other academic position	4
K-12 education position	1
Non-profit organization	1
Government	4
Industry	24
Self employed	7
Other position	4
Total	100%

Finally, to ascertain the breadth of career options that science and mathematics Ph.D.s entertain, we asked respondents to indicate all careers they have ever considered. On average, each respondent has considered four different career paths, although a few have considered as many as eight. Similar percentages have considered a position in industry (73 percent) and a position as faculty in a research university (70 percent). Half of the respondents have considered government employment (52 percent) and a quarter (28 percent) have considered working with non-profit organizations. Interestingly, a position in K-12 education was a career path considered by 26 percent of the respondents (N=168). The distribution of long-term careers ever considered is shown in Table 2-3.

TABLE 2-3 Percentage Distribution of Positions Ever Considered

Position Ever Considered	Percent of all respondents (N=680)
Postdoctoral position	38%
Research university faculty position	70
Liberal arts college position	43
Community college faculty	27
Other academic position	42
K-12 education position	26
Non-profit organization	28
Government	52
Industry	73
Self employed	22
Other position	3

INTEREST IN SECONDARY SCHOOL EDUCATION

Defining Interest in Secondary School Education

There were three parts of the questionnaire that identified respondents' interest in a position in K-12 teaching or education. One question asked if the respondents had ever "considered a position in K-12 science/ mathematics education, teaching, or leadership." This response was included among the many alternatives to the questions on next employment step, career aspirations and positions ever considered. One-quarter of the respondents (N = 168) indicated that they had considered a position in K-12 education. In addition, respondents were asked "Have you ever actively considered taking a position teaching science or mathematics in a secondary school (grades 7-12)?" Just over 30 percent (N = 211) of respondents indicated that they had actively considered a career in secondary school teaching.

In answering these two questions, a total of 245 respondents (36 percent) indicated that they had considered either a position in K-12 education or a secondary school teaching position.[1] Conversely, 82 (12 percent) respondents indicated that they would not consider any secondary school position under any circumstances. Finally, 353 respondents (52 percent) fell into neither of these two groups and were classified as having no prior consideration of a position in secondary school education as a career alternative. Based on these responses, we classified respondents on the basis of their interest in or consideration of a position in secondary school education. A total of 245

[1] For some unexplained reason, 77 survey respondents who indicated that they had actively considered secondary school teaching, did not indicate that they had aspired to a career in K-12 education.

respondents had considered and not rejected a position in secondary school education; 353 had never thought about a secondary school position; and 82 had considered and rejected a secondary school position. In terms of their interest in a position in secondary school education the respondents are categorized as *yes,* there is an interest in secondary school education (N=246); they had *not considered* a position in secondary school education (N=353); and there is *no interest* in secondary school education (N=82).

The respondents' interest in careers in secondary school education varied according to a number of social and demographic variables. For example, graduate students were somewhat more likely to have considered a position in secondary school education than were postdocs. Nearly 37 percent of graduate students indicated having considered a career in secondary school education compared to 28 percent for postdoctoral fellows. However, 51 percent of respondents classified as "other" (Ph.D.s not in postdoctoral positions) had considered a career in secondary school education. Younger respondents were slightly more likely to have considered a career in secondary school education. Just over 38 percent of respondents under age 30 indicated they had considered a career in secondary school education compared to 34 percent for respondents age 30 and over.

There was also variation in whether the respondent had considered a career in secondary school education by the respondent's discipline. Respondents in mathematics and physics were most likely to have considered a career in secondary school education (44 percent) followed by those in biological sciences (37 percent), earth sciences (30 percent) and chemistry (27 percent).

There was a significant difference in the percentage of respondents that had considered a secondary school position by gender ($p < .01$). A higher percentage of female respondents (43 percent) had considered a secondary school position than male respondents (32 percent).

Citizenship was also significantly associated with consideration of a secondary school position ($p < .005$). As might be expected, 41 percent of U.S. citizens (native and naturalized) had considered a position in secondary school education, compared to only 27 percent of temporary and permanent residents. Although, respondents who identified themselves as white (38 percent) were more likely to have considered a position in secondary school education than those respondents of other ethnic origins (33 percent), the difference was not significant.

There was no association between family status and prior consideration of a position in secondary school education. Neither marital status nor the presence of children in the household was associated with prior consideration of a position in secondary school education.

All but eight of all respondents had some prior teaching experience and many respondents had multiple forms of teaching experience. Most (87 percent) have served as teaching assistants. In addition, about 10 percent of the respondents had experience teaching in either elementary or secondary school. About 22 percent had teaching experience through volunteer work in secondary schools.

Financial Considerations

To understand better how financial considerations might affect career choices, we asked respondents to provide us with their current salary, the salary they expected at their immediate next step of employment, and the salary they expected seven years from now when, presumably, they would have settled into a stable career. We also asked respondents to provide us with the beginning salary they would require if they were to assume a position as a secondary school teacher and the salary they would require after seven years in secondary school teaching.

The average current salary of respondents, most of whom are graduate students, was about $19,661. The current salary of postdoctoral fellows, $28,667, was significantly higher that the current salary of graduate students, $15,972 ($p < .001$).

The respondents anticipated that at their immediate next step of employment their average income would be $37,753. About half of the respondents expected to hold postdoctoral fellowships as their next step in employment and they anticipated, on average, a salary of $29,500. For respondents planning on faculty positions at a research university, the average anticipated salary was $45,642. For those who planned a next step as faculty at a liberal arts college, the anticipated average salary was $40,259. Finally, respondents entering into industry expected a beginning salary of $52,082.

Respondents were asked to anticipate their salaries in seven years, presumably at a time when they would all be in a stable career in the positions to which they aspired. The average salary for all respondents seven years out was $71,627 (see Table 2-4). Those respondents who sought self-employment anticipated the highest salaries in seven years, $121,276. Those planning careers in industry anticipated salaries of $80,812 seven years out. Respondents who planned careers in academia anticipated salaries that were considerably lower. For example, the salary anticipated in seven years by faculty in research universities was $65,819, and for faculty in liberal arts colleges only $56,113.

TABLE 2-4 Respondent's Anticipated Average Salary in Seven Years by Ultimate Career Aspiration

Ultimate Career Aspiration of All Respondents	Anticipated Salary in Seven Years
Research university faculty position	$65,819
Liberal position arts college	56,113
Community college faculty	37,116
Other academic position	54,222
K-12 education position	54,375
Non-profit organization	60,000
Government	73,692
Industry	80,812
Self employed	121,276
Average, all respondents	$71,627

All respondents were asked what they would require as a starting salary in order to consider a secondary school teaching position. The average of $37,128 is not very different from the $37,752 all respondents anticipated at the next step in their employment and considerably larger than the salary expected by respondents who chose to be postdocs at their next step. In addition, all respondents were asked for the salary they would require after seven years in order to consider secondary school teaching. The average salary of $52,828 is considerably lower than the average of $71,627 all respondents anticipated in their careers seven years from now (see Table 2-5).

There were no significant differences in either anticipated salary at the next step of employment or salary anticipated in seven years among respondents based on their prior interest in secondary school education. However, respondents who had demonstrated some prior interest in a position in secondary school education had significantly lower salary expectations *for a position in secondary education* than did those who had not considered or were not interested in a position in secondary school teaching.

TABLE 2-5 Beginning Salary and Salary in Seven Years Required to Consider a Career in Secondary School Teaching by Interest in a Position in Secondary School Education

Interest in Secondary School Education	Average Starting Salary[1]	Average Salary in Seven Years[1]
Yes (N=245)	$35,795	$52,828
Not Considered (N=353)	37,460	55,973
No Interest (N=82)	41,037	61,370
Total (N=680)	$37,128	$55,216

[1] $p < .001$

ASPECTS OF A PROGRAM TO ATTRACT SCIENCE AND MATHEMATICS PH.D.S TO SECONDARY SCHOOL EDUCATION

To understand better the kinds of structural changes that might facilitate the entry of science and mathematics Ph.D.s into secondary school education careers, we asked graduate students and Ph.D.s in our sample to respond to a number of scenarios in which they were asked to indicate if they would consider accepting a position as a secondary school science or mathematics teacher. These scenarios dealt with a number of perquisites suggested by the focus group discussions that might be provided to Ph.D.s engaged in secondary school teaching. The perquisites that were tested included a national fellowship, a guaranteed two-year postdoctoral fellowship following two years of secondary school teaching, the option of teaching in a specialized science and technology high school, financial assistance for education expenses, student loan forgiveness, and teaching resource support. Respondent's responses to these prerequisites are shown in Table 2-6.

TABLE 2-6 Percentage of Respondents Agreeing to Conditions for a Position as a Secondary School Science or Mathematics Teacher by Prior Interest in a Career Position in Secondary School Education

I Would Consider Taking a Position as a Secondary School Science and Mathematics Teacher if....	Interest in Position in Secondary School Education			All Respondents (N=680)
	Yes (N=245)	Not Considered (N=353)	No Interest (N=82)	
I were awarded a prestigious national fellowship that provided training, placement, and covered living expenses	87%	66%	12%	67%
I had the option of 2-year postdoc following a 2-year teaching commitment, in addition to the national fellowship	73	67	11	63
I had the option of 2-year postdoc following a 2-year teaching commitment without the option of the national fellowship	57	37	7	41
I were able to teach in a specialized science and technology high school	81	48	13	56
I received support from a regional or university-based science resource center that provided science kits and loaned equipment	75	44	12	52
Financial assistance were available for the education expenses related to the certification process	40	11	8	21
One year of student loans were forgiven for each year of K-12 teaching	57	38	11	42

As expected, respondents who had previously considered a position in secondary school education were much more likely to consider secondary school teaching positions under all of these scenarios. Only a handful of those respondents with no interest in a position in secondary school education would consider secondary school teaching under any scenario. Generally speaking, respondents who had not previously considered a position in secondary school education were similar to those who had. Some respondents who had considered a position in secondary school education would not consider teaching under any scenario, and some of the respondents who had no interest in secondary school education would consider each scenario.

Respondents who had previously considered a position in secondary school education endorsed four of the scenarios. These include:

- The award of a prestigious national fellowship that would provide training, placement, and special opportunities for networking with peers, as well as cover living expenses during the training period (87 percent).
- The option of a two-year guaranteed postdoctoral research fellowship at the end of a two-year teaching commitment, in addition to the national fellowship described above (73 percent).
- The ability to teach in a specialized science and technology high school (81 percent).
- Support during teaching through a regional or university-based science teaching resource center that provided science kits, loaned laboratory equipment, and organized field opportunities for science experiments in which students could participate (75 percent).

These scenarios were also endorsed by respondents who had not previously considered a position in secondary school education, but not to the same degree. Two-thirds of respondents in this category endorsed the award of a national fellowship and the option of a two-year guaranteed postdoc in addition to the fellowship. Nearly half of these respondents favored teaching in a science and technology high school and the support of a science resource center.

Participants who had previously considered a position in secondary school education were less enthusiastic about the remaining three scenarios. Two scenarios were endorsed by more than half of these respondents and one was endorsed by only 40 percent of respondents. These scenarios include:

- The option of a two-year guaranteed postdoctoral research fellowship at the end of a two-year teaching commitment without the national fellowship (57 percent).
- One year of my student loans forgiven for each year of employment in a full-time teaching position (57 percent).
- Financial assistance made available for the educational expenses incurred in the process of gaining certification (40 percent).

Respondents who previously had not considered a position in secondary school education were much less likely to endorse these three scenarios. Only 37 percent endorsed the two-year postdoc without the national fellowship; 38 percent endorsed the forgiveness of student loans; and only 11 percent were endorsed financial assistance for certification expenses.

Relief of education debt was endorsed by 42 percent of all respondents. Over half (57 percent) of the respondents who had previously considered a position in secondary school education were interested in education debt relief compared to 38 percent for those respondents who had not previously considered a position in secondary school education and only 11 percent of those with no interest in secondary school education. These differences may be the result of differences in the level of education debt among the three groups. Respondents who had previously considered a position in secondary school teaching had higher levels of educational debt than did other respondents, although this difference was not statistically significant. Less than half (48 percent) of respondents reported any educational debt. For respondents with educational debt, the average level of debt was $20,817. Respondents who endorsed education debt relief had a significantly higher level of education debt, $23,585, than did those who did not endorse debt relief, $15,852 ($p < .001$).

As part of the process for developing our survey questionnaire, we held a series of focus groups with graduate students and postdoctoral fellows on their career aspirations and interest in a position in secondary school education. Participants in the focus groups told us that there were five additional issues that would potentially affect whether some of them would consider a career in secondary school science and mathematics teaching. We developed a series of scaled-response questions to assess the importance of these issues for survey respondents in their consideration of a position in secondary school education. These five issues were:

(1) The availability of funded fellowships for work in research laboratories during the summer months;
(2) The ability to get release time and travel expenses during the school year to attend professional meetings;
(3) The availability of experienced teachers to serve as mentors during the school year;
(4) The ability to receive teacher certification through some expedited process; and

(5) A time-limited commitment to secondary school teaching.

The graduate students and postdoctoral fellows in the survey were asked to consider a number of options for each of the conditions and respond yes or no to whether these options would affect their consideration of secondary school teaching. Each respondent was able to select more than one response. Consequently, the number of positive responses is greater than the number of respondents. Some respondents, especially those with no interest in teaching, selected none of the responses.

As in the analysis of the previous scenarios, we have used the classification of respondents by their level of interest in secondary school education in this analysis. Respondents who had previously expressed an interest in secondary school education were most likely to indicate some level of consideration of conditions for employment in secondary school teaching. Respondents who had not previously considered a position in secondary school education were less likely to consider alternative conditions for secondary school teaching. Generally speaking, these were differences in degree between the response patterns of respondents who had considered secondary school education and those who had not. However, those who had no interest in secondary school education responded in a very different pattern. The following tables present the responses to each specific scenario.

Summer Research. As seen in Table 2-7, many respondents consider a guaranteed position in a research lab for the summer months an important prerequisite in considering a career in secondary school teaching. On the other hand, a number of respondents indicated that freedom to do as they pleased during the summer would make secondary school teaching attractive. A majority of respondents who had expressed an interest in secondary school education would consider a secondary school teaching position if they were guaranteed a summer fellowship in a research lab, including travel expenses. This was also true of those respondents who had not previously considered a position in secondary school education. A guaranteed summer fellowship was of little interest to respondents who indicated no interest in secondary school education.

TABLE 2-7 Percentage of Respondents Attracted to Secondary School Teaching by Availability of Summer Research and Interest in Secondary School Education

I Would Consider Taking a Position as a Secondary School Science and Mathematics Teacher if....	Interest in Position in Secondary School Education			All Respondents (N=680)
	Yes (N = 245)	Not Considered (N = 353)	No Interest (N = 82)	
I were guaranteed summer fellowship in a research lab including travel expenses	79%	63%	17%	63%
I were guaranteed summer fellowship in a research lab	63	42	11	47
I had the opportunity to compete for summer fellowship in a research lab	41	18	6	25
I had the opportunity to participate without compensation in a lab during the summer	21	6	1	11
I were free to do as I pleased during the summer	62	26	12	37
No interest in teaching under any of these scenarios	7	30	80	28

Participation at Scientific Meetings. The respondents indicated that both release time and funding to attend at least one professional meeting during the school year would be an important factor in considering a position in secondary school teaching. As shown in Table 2-8, there is a considerable reduction in positive responses when the condition is restricted to just release time to attend professional meetings without funding.

TABLE 2-8 Percentage of Respondents Attracted to Secondary School Teaching by Participation at Scientific Meetings and Interest in Secondary School Education

I Would Consider Taking a Position as a Secondary School Science and Mathematics Teacher if....	Interest in Position in Secondary School Education			All Respondents (N=680)
	Yes (N = 245)	Not Considered (N = 353)	No Interest (N = 82)	
I were given funding and time to attend two or more scientific meetings during the school year	88%	61%	17%	63%
I were given funding and time to attend one scientific meeting during the school year	77	47	12	53
I were given time to attend scientific meetings during the school year	48	24	7	31
I were given time to attend one scientific meeting during the school year	39	14	5	22
I would have no opportunity to attend scientific meetings during the school year	13	2	5	6
No interest in teaching under any of these scenarios	7	30	80	28

Assistance in Curriculum Development and Teaching. As seen in Table 2-9, respondents indicated that the assistance of an experienced teacher to serve as a mentor was an important factor in considering secondary school teaching. It is interesting to note that the duration of the assistance from an experienced teacher is not as important as the availability of assistance.

TABLE 2-9 Percentage of Respondents Attracted to Secondary School Teaching by Interest in Mentoring and Interest in Secondary School Education

I Would Consider Taking a Position as a Secondary School Science and Mathematics Teacher if....	Interest in Position in Secondary School Education			All Respondents
	Yes (N = 245)	Not Considered (N = 353)	No Interest (N = 82)	
I were mentored by a selected, experienced teacher during my initial two years of teaching	60%	35%	7%	40%
I were mentored by a selected, experienced teacher during my initial year of teaching	68	41	6	47
I were mentored by a selected, experienced teacher during my initial semester of teaching	60	32	11	40
I were mentored by a selected, experienced teacher on an as-needed basis	69	34	12	44
I would have no formal assistance from a mentor	27	12	5	17
No interest in teaching under any of these scenarios	6	33	82	29

Teacher Certification Requirements. Respondents clearly indicated that they were unwilling to spend more time than an intensive summer session to attain teacher certification. A majority of respondents indicated that teaching proficiency demonstrated on-the-job should suffice for teacher certification (see Table 2-10). However, a substantial proportion of respondents indicated that they would be willing to attend an intensive summer course in education in order to obtain teacher certification. Few respondents were willing to spend any additional time in coursework leading to teacher certification. Generally speaking, respondents who had previously considered a position in secondary school education were more willing to commit to additional training for teacher certification.

TABLE 2-10 Percentage of Respondents Attracted to Secondary School teaching by Procedures for Teacher Certification and Interest in Secondary School Education

I Would Consider Taking a Position as a Secondary School Science and Mathematics Teacher if....	Interest in Position in Secondary School Education			All Respondents (N=680)
	Yes (N = 245)	Not Considered (N = 353)	No Interest (N = 82)	
Teaching proficiencies could be demonstrated on the job	80%	61%	16%	63%
I could demonstrate proficiency prior to becoming a teacher through prescribed formal courses or experiences based on my needs	69	41	11	48
I could receive teacher certification by taking intensive summer courses in education	68	35	7	44
I could receive teacher certification by taking one year of education courses	22	10	6	14
I could receive teacher certification by taking more than one year of education courses	6	3	2	4
No interest in teaching under any of these scenarios	5	26	82	26

Duration of Teaching Commitment. Respondents were asked to indicate how long a commitment they were willing to make to teaching in a secondary school (see Table 2-11). For respondents who both had and had not previously considered a position in secondary school education, any commitment longer than two years was not conducive to considering secondary school teaching.

TABLE 2-11 Percentage of Respondents Attracted to Secondary School Teaching by Duration of Teaching Commitment and Interest in Secondary School Education

I Would Consider Taking a Position as a Secondary School Science and Mathematics Teacher if....	Interest in Position in Secondary School Education			All Respondents (N=680)
	Yes (N = 245)	Not Considered (N = 353)	No Interest (N = 82)	
I would accept a 1-year teaching commitment	79%	55%	15%	58%
I would accept a 2-year teaching commitment	73	42	7	49
I would accept a 3-year teaching commitment	42	17	5	29
I would accept a 4-year teaching commitment	24	7	4	13
I would accept a 5-year teaching commitment	24	8	4	13
No interest in teaching under any of these scenarios	4	29	82	27

INTEREST IN OTHER K-12 EDUCATION POSITIONS

In the course of holding focus group discussions in preparation for developing its survey questionnaire, the committee found that many participants believed there were a number of ways, beyond secondary school teaching and curriculum development, that Ph.D.s could contribute to improving science and mathematics education. Graduate students and postdoctoral fellows who participated in focus group sessions clearly liked the idea of working on the development of science and mathematics curricula, not just for secondary school courses but across grades from kindergarten to grade 12. Further, focus group participants suggested that Ph.D.s could also contribute by working in professional development, science education partnerships, or science museums and environmental centers. They also indicated that Ph.D.s could contribute as science or math specialists in school districts, science resource teachers in elementary or secondary schools, science specialists in regional or university-based science education resource centers, or K-6 science or math teachers. The committee believed these suggestions were important to explore both in their own right and as context for the more specific questions on secondary school teaching and curriculum development it was asked to address. The suggestions of focus group recipients were addressed by adding one question to the survey questionnaire.

Respondents were asked to indicate if they would consider full-time employment in a wide range of K-12 education positions other than secondary school teaching (see Table 2-12). Respondents were able to identify up to eight K-12 education positions. There was a very high level of interest among respondents in these positions; the typical respondent indicated an interest in four different K-12 education positions. The most popular K-12 education positions were:

- Professional development
- Science or mathematics specialist
- University or industry-based science educational partnerships
- Science specialist in a science resource center

These positions were considered by more than three-quarters of respondents who had a prior interest in K-12 education and by more than half of those who had not previously considered K-12 education careers.

More than half of all respondents indicated an interest in two other positions. Respondents who had previously considered a position in K-12 education were considerably more interested in these positions than other respondents. These positions include:

- Curriculum development
- Working in a science museum or similar institution

Finally, fewer than half of the respondents indicated interest in two additional positions:

- Science resource teacher
- K-6 science or mathematics teacher

TABLE 2-12 Percentage of Respondents Attracted to Positions in K-12 Science or Mathematics Education by Interest in Secondary School Education

Secondary School Science or Mathematics Education Position	Interest in Position in Secondary School Education			All Respondents (N=680)
	Yes (N=245)	Not Considered (N=353)	No Interest (N=82)	
Curriculum development (devising or refining units of study by grade level)	66%	37%	16%	55%
Professional development (teaching science or mathematics teachers)	82	54	23	74
Science or mathematics specialist for a school district	80	53	17	71
Science resource teacher for an elementary or secondary school	55	28	12	44
K-6 science or mathematics teacher	26	11	6	19
Science Education Partnerships (university or industry-based expertise for K-12 classrooms)	75	55	25	72
Science specialist in a regional or university-based science education resource center	74	58	28	73
Working in a science museum, environmental science center, or similar institution	66	49	28	64

While it would have been difficult to test for this through our survey, we did note in holding focus group discussions that the individuals who were interested in such positions as curriculum development specialist recognized that becoming expert in curriculum development took years of experience in the classroom. Indeed, it was clear that those who would be most interested in curriculum development would seek substantial classroom experience first.

SUMMARY

Level of Interest

In 1997, only 0.8 percent of persons with Ph.D.s in the physical, biological, or mathematical sciences held a position in K-12 educational institutions. This survey demonstrates, however, that a surprisingly large percentage of science and mathematics graduate students and recent Ph.D.s among our respondents have considered a position in secondary school education. Respondents who were younger, female, and native born were the most willing to consider a position in secondary school education.

Potential Incentives

The survey also demonstrates that there are certain conditions under which Ph.D.s might potentially find careers in secondary education more attractive. We summarize these conditions here, and discuss our judgments about the pros and cons of implementing any one of them in the conclusions to Chapter 4.

In sum, then, respondents were presented with a number of scenarios that asked whether they would consider taking a position as a secondary school teacher in science or mathematics. In one set of questions, four scenarios emerged that showed promise for attracting Ph.D.s to secondary school careers. These were:

(1) The award of a prestigious national fellowship that would provide training, placement, and special opportunities for networking with peers, as well as cover living expenses during the training period;
(2) The ability to teach in a specialized science and technology high school;
(3) Support during teaching through a regional or university-based science teaching resource center that provided science kits, loaned laboratory equipment, and organized field opportunities for science experiments in which students could participate; and
(4) The option of a two-year guaranteed postdoctoral research fellowship at the end of a two-year teaching commitment, in addition to the prestigious national fellowship.

There were other issues that also had an impact on whether respondents would consider taking a position as a secondary school science or mathematics teacher. Two of these issues dealt with maintaining linkages to colleagues and updating scientific skills. First, the availability of funded summer fellowships in research labs, with travel expenses, during the summer was an important factor in whether respondents would consider a secondary school teaching position, selected by nearly 80 percent of respondents who had previously considered a position in secondary school teaching and 63 percent of all respondents. Second, 88 percent of respondents who had previously considered a position in secondary school teaching and 63 percent of all respondents indicated they would consider a position as a secondary school science or mathematics teacher if funding and release time to attend two scientific meetings during the school year were made available.

Respondents recognized that by teaching in a secondary school system, they would encounter barriers. Two issues of concern were teacher certification and acquiring skills in pedagogy appropriate to secondary schools. While most respondents preferred teacher certification based on proficiencies demonstrated on-the-job, a majority would consider participating in formal coursework in education during an intensive summer course to obtain certification. Two-thirds of the respondents who had previously considered a position in secondary school education and 54 percent of all respondents wanted some level of mentoring from an experienced science teacher, but there was little consensus on the optimal amount of mentoring. Finally, most respondents indicated that they were unwilling to commit initially to more than two years of teaching mathematics or science in a secondary school.

Respondents who had previously considered a position in secondary school education had lower salary expectations for a secondary school teaching position, both for beginning salaries and salaries anticipated in 7 years. However, all respondents recognized that salaries for positions in secondary teaching were lower than for other career alternatives available to them.

Other K-12 Positions

Finally, respondents expressed considerable interest in a wide range of K-12 education positions other than secondary school teaching. Positions considered by more than half of respondents who had previously considered a position in secondary school teaching, as well as who had not considered such a position, include professional development, science or mathematics specialist for a school district, university or industry-based science educational partnerships, and science specialist in a science resource center.

3

Perspectives on Ph.D.s in Secondary School Science and Mathematics Education

To complement the information we collected on the interest of Ph.D.s in secondary school education careers, we sought to learn from the experiences and perceptions of individuals already in the field on the opportunities and obstacles for Ph.D.s in secondary school education careers. First, we interviewed eighteen Ph.D.s working in secondary school education, mostly as secondary school teachers, but some in other positions. The object of these interviews was to examine how they made the transition to the secondary school environment, their challenges and successes, and their recommendations for how other Ph.D.s could contribute to secondary school education.

We also interviewed five high school principals, three school district superintendents, and four state school officials. The high school principals included those in charge of regular public schools, as well as science and technology magnet schools. These administrators work in New Jersey, North Carolina, Texas, California, and Washington. The object of the interviews was to understand the perspectives of educators on the contributions Ph.D.s might be able to make in secondary school education and the conditions that might influence their success. We also interviewed five graduate deans to explore how graduate students might be better prepared for secondary school careers and to learn about opportunities for restructuring programs to support those who might seek this career option.

In the course of our interviews, we encountered a number of stereotypes about science and mathematics Ph.D.s and the secondary school educational system. We also found some surprises about the experiences of those Ph.D.s who have taken positions in secondary school education and creative suggestions that could make secondary school positions more viable, fulfilling, and thus more attractive to scientists. Appendix A discusses our study methodology in detail, including our process for identifying those interviewed. Interview guides can be found in Appendix C. Detailed results of our interviews with Ph.D.s can be found in Appendix D and results of our interviews with administrators in Appendix E. The individuals we interviewed cannot be considered a representative sample, but the discussions raise many important issues and ideas that should be discussed further at the national workshop to be held in the spring of 2000 as part of the next phase of this project.

PH.D.S IN SECONDARY SCHOOL SCIENCE AND MATHEMATICS EDUCATION

In our discussions with science and mathematics Ph.D.s and administrators we sought to assess both perceptions and realities about the strengths and weaknesses of Ph.D.s in secondary school science and mathematics education generally and in secondary school science and mathematics teaching in particular.

In general, we were struck by a pattern of perceptions among administrators about how Ph.D.s would fare in secondary school teaching. Administrators agreed that multiple obstacles keep a large number of Ph.D.s from considering positions in secondary school education. Almost all of the administrators cited low salaries as a one obstacle and the potential opposition of teacher unions to recruitment of Ph.D.s as another. Several believed that Ph.D.s would encounter difficulties with certain aspects of the secondary school teaching environment, like having to deal with bureaucracy, receiving little respect, having to discipline students, and working with students who say "why do I need to learn this?" High school and magnet school principals saw several additional drawbacks to science Ph.D.s in secondary school classrooms. First, they tended to see Ph.D.s as overspecialized and overprepared for teaching secondary school students. Along with district superintendents, they were also more likely than state officials and graduate deans to state that Ph.D.s needed educational coursework—and certification— before teaching.

School administrators identified a number of strengths that Ph.D.s could bring to secondary school science and mathematics positions. They cited most often the content knowledge of the Ph.D. and argued that the experience of Ph.D.s in scientific inquiry provides them an important skill that they could bring to problem-based teaching. Those interviewed also cited the ability of Ph.D.s to assist in the implementation of national science and mathematics education standards. Administrators saw Ph.D.s as contributing organizationally as well. Both magnet and regular high school principals saw Ph.D.s as enhancing their schools' credibility and standing with accrediting bodies and as individuals who could obtain grant funding for school programs. Administrators also saw the potential for Ph.D.s to assist in building links between high schools, colleges, and universities. Some administrators suggested that Ph.D.s could bring a positive attitude and energy that would be valuable in the classroom.

The Ph.D. teachers we spoke with claimed they were also more flexible in their teaching styles than other teachers. They saw themselves as less dependent on textbooks for exercises and more able to create challenging environments in the classroom through the variety of ways they could present material. Ph.D. teachers saw a role for others like themselves in problem-based inquiry learning programs. In their view, Ph.D.s excel as teachers in such programs because they know the content and can concentrate on the delivery and interpretation of data with the students. Frequently, teachers without Ph.D.s do not use this style of teaching because they are afraid to tell their students "*I don't know*" when they get unexpected results. Ph.D.s have significant experience in dealing with unexpected results and formulating new hypotheses. Two-thirds of the Ph.D. teachers believed that the future for Ph.D.s in science teaching was "good" to "very good," especially in light of the new science curriculum reform movement. They felt more teachers were needed in this area, and they had to be better trained. Ph.D. teachers offered the opinion that other Ph.D.s interested in secondary school teaching should volunteer in secondary schools both to help science teachers in inquiry-based learning and to obtain classroom experience.

WHICH PH.D.S?

An important lesson learned from discussions with Ph.D.s working in secondary school education is that many generalizations about Ph.D.s are irrelevant to a discussion of how Ph.D.s might contribute to secondary schools. Most individuals who have pursued and obtained a Ph.D. degree in science or mathematics are specifically interested in research and, typically, have personalities suited to the research environment. But not everyone fits the stereotype of the laboratory scientist or abstracted mathematician. The majority of Ph.D.s in secondary school classrooms that we interviewed said they are content or very happy with their careers in secondary school education. A frequent comment was that they had found research isolating and were much happier in a job that required more interaction with people.

Teaching appears to be a viable alternative for Ph.D. scientists with certain characteristics. First, they must enjoy interacting with others, and especially with students. The Ph.D.s we spoke with said they enjoyed working with school-age children and found that helping students learn and succeed is fulfilling. Second, those who succeed love teaching—the enjoyment of teaching outweighs its potential negatives, such as lower salaries than in other science and mathematics careers. An important piece of advice Ph.D. teachers had for those contemplating a career in secondary school education was to assess their own personalities to determine if they like working with people in a setting very different from the academic or industrial research environment.

We discovered that lower salaries were less of an issue for the group of Ph.D.s we spoke to than expected. Many of the Ph.D.s who decided to pursue teaching careers had family or geographic restrictions related to their spouses' jobs. However, many were dependent on their spouses' income for a sense of financial security. Others were teaching while in "retirement" from previous jobs. Teaching was a "luxury" available to them because of these other sources of income.

Finally, several Ph.D.s indicated that they felt that they were giving something back to their communities and that this, too, was rewarding. As one Ph.D. put it: "I'm not so excited about creating a billion scientists, but science literacy is something I really believe in, as an equalizer or way for people to take care of themselves, learn about the world, make sure they are not taken advantage of, and preserve the environment...those are all important things that you can address best with K-12 education."

PREPARATION FOR TEACHING

High school and magnet school principals uniformly agreed that Ph.D.s would not be able to teach effectively without educational coursework. Graduate students and postdoctoral fellows we spoke to in focus groups, however, perceived the process of taking education courses and obtaining teacher certification as major obstacles to becoming teachers. They saw little value in education courses, large costs in wages foregone in taking them, and the entire process as taking too much time.

Science and mathematics Ph.D.s teaching in secondary schools offered what we believe are insights and suggestions about preparing Ph.D.s for teaching that may bridge the divide between the requirements of school systems on the one hand and the perceptions of graduate students and recent Ph.D.s on the other. Several science and mathematics Ph.D.s recommended that graduate students or recent Ph.D.s considering teaching should obtain certification even if they plan to work at a private school. They believe that there is value in psychology and pedagogy courses and noted that the teaching certificate provides one with a wider range of options. However, they argued that there should be an alternative certification process available for those with Ph.D.s in science and mathematics. This process should be accelerated and subsidized so that Ph.D.s could afford to do it during a summer prior to teaching or during a semester prior to the completion of their Ph.D.s. They also argued that the content of the teacher education courses that Ph.D.s would take should build on their experience and target their specific needs.

PRACTICAL ASPECTS OF THE TRANSITION TO THE SECONDARY SCHOOL ENVIRONMENT

We asked those we interviewed about the practical steps a graduate student or Ph.D. needs to take to move into a secondary school environment and found a consensus across all groups that anyone who is considering teaching as a career needs experience in a classroom with students. Ph.D. teachers strongly suggested that if a Ph.D. candidate is seriously interested in teaching, that person should volunteer in a classroom, substitute teach, or seek other means for obtaining classroom experience during graduate school, so as to make an informed career decision. They also argued it is important to have more opportunities like the National Science Foundation's National Graduate Teaching Fellows in K-12 Education program (popularly known as the GK-12 program), that supports graduate students who want to improve their teaching skills while finishing their Ph.D.s.

Principals, superintendents, and graduate school deans agreed that Ph.D.s should have some experience with classrooms and students, especially while in graduate school. At least one individual within each group of administrators suggested Ph.D.s should be provided experience working in classrooms/schools during graduate school. In a similar vein, at least one individual within each group of administrators (except magnet school principals) indicated Ph.D.s should take educational coursework while pursuing their advanced degrees.

All of the administrators were uncertain whether graduate schools would be willing or able to support graduate students in the sciences who want to explore secondary school education—the typical response was "maybe." Many administrators suggested that improved links between education schools and schools of arts and sciences (science and mathematics departments) within colleges and universities would improve science and mathematics teacher education. They argued that teacher preparation should include more subject matter courses. They also argued that faculty in arts and sciences underestimate the value education courses add for science and mathematics students considering teaching.

We asked Ph.D. teachers what they encountered when they told others in the university environment that they had decided to pursue a secondary school teaching career and found that there had been a surprising amount of support for these individuals. One-third of the Ph.D.s did relate that their decisions were met with "total disbelief," and several reported that others told them they would be wasting their education. One teacher related that a professor told her that "we are not training our students…to be high school teachers. We are investing $100,000 per student…they need to return that investment to the scientific community." Another related that colleagues who saw her research career as progressing well asked her "why she wanted to *stoop* to high school teaching." However, one-quarter said that their decisions to pursue a secondary school career were met with almost no reaction, and half of the Ph.D.s said that their mentors and colleagues were generally supportive.

Several Ph.D.s indicated that changing careers is not as difficult as it is often perceived to be. One noted that the people who are "leaving science are not the ones who were bad scientists. Too many people stay around because they don't have a clue as to what else they can do." In the same vein, another (who had returned to full time research) noted "I am not afraid of switching careers…I will try teaching again if I get bored with the work at the [research] institute. It is really hard to change careers the first time, but after that it is no big deal." One respondent noted there was a group of graduate students at a major research university who felt that they were failures in the eyes of their peers. These students "found research is cold and impersonal, but didn't know how to break through without starving to death. I just said do it, have some faith in themselves. After all, they have half the battle taken care of—they know the content, they have time to focus on their people skills."

The Ph.D. teachers we talked to suggested that if, prior to getting the master's degree, graduate students realized that they dislike research and really want to teach, they should stop at the master's degree and shift into teaching right away. For individuals who already have the master's degree, however, interviewees recommended that they complete the Ph.D. because it would help open some doors in the future. When asked whether he was glad he had finished the Ph.D., one teacher said "I was really glad to have done the intense work, so I could concentrate on the teaching and model being a scientist. You can afford to have fun!"

PH.D.S IN THE SECONDARY SCHOOL WORK ENVIRONMENT

Another common perception about Ph.D.s is that they would have trouble feeling at home in the secondary school work environment. Our conversations with Ph.D.s and administrators suggested that, to the extent this is a problem, creativity and resources could address these issues.

There is a common misperception that Ph.D.s would not be made welcome by other secondary school teachers. Most of the Ph.D. teachers indicated that the Ph.D. was not an issue for their secondary school colleagues. A handful noted some initial resistance or caution among colleagues but that this evaporated once the Ph.D.s had proven themselves as good teachers. One Ph.D. acknowledged that if one comes across as "pontificating or bossy," then having a Ph.D. would be a hindrance to forming good working relationships with others; another noted that personality does play an important role in forging relationships in the workplace and that "a Ph.D. who goes into teaching is not typically arrogant."

Similarly, Ph.D. teachers did not meet the level of resistance from school administrators that is commonly perceived to exist. Only about one-third of Ph.D.s said they had a difficult time in the job interviews convincing school administrators that they were serious about their desire to teach, and this was a real obstacle for them. One stated: "Most people don't trust Ph.D.s...they think they won't be able to relate to other people, that they won't be a team member. Also a lot of people are afraid of being upstaged." That individual encountered the sentiment that "why in the world would you take a Ph.D. and teach high school...you must be a failure." Several Ph.D.s got their first positions only because the school districts that hired them had positions they could not fill. They noted that subsequent jobs were easier to land once they had proven they were committed to this career path. Moreover, one-third of the Ph.D. teachers also noted that once hired, their school administrators saw them as an asset and often noted publicly that they had a Ph.D. on staff.

There is a common perception, particularly among administrators, that in order to attract Ph.D.s to secondary school teaching, changes need to be made to "make it interesting" for them and to support them. As we have already discussed, most Ph.D.s who pursue secondary school teaching are—and should be—individuals who love to teach and enjoy working with students first of all. Having said this, both Ph.D.s and administrators put forward a number of suggestions that should be seriously considered.

They suggested that programs allowing Ph.D.s to engage in research during the summer months would benefit them personally and as teachers. One state official noted that programs already exist there that bring high school teachers into industrial laboratories during the summer months. Linking Ph.D. science teachers to these programs should be an important goal of any program designed to attract Ph.D.s to secondary school teaching. This could be beneficial in several ways: (1) Ph.D.s would continue to feel connected to scientific research, (2) they could bring their summer experiences back to their teaching during the school year, and (3) these opportunities could potentially provide 12-month instead of 9-month employment for Ph.D.s and, therefore, higher overall salaries.

The Ph.D. scientists made similar assertions. They encouraged the development of long term relationships between Ph.D. teachers and research laboratories (academic or industrial). Ph.D.s could return every summer to research laboratories for ten to twelve weeks to do research. They suggested that Ph.D.s could focus on a particular component of a research project; Ph.D.s would likely be of greater value than a high school, college or graduate student; and Ph.D.s would be more likely to return in subsequent summers. In addition, laboratory scientists with whom the Ph.D. teachers worked during the summer could be a potential resource during the school year for mentoring students or providing surplus laboratory supplies.

In an interesting variation on this idea, one person recommended that cyclical industries such as petroleum and aerospace look into supporting their Ph.D. employees who want to get into secondary school teaching as a counter-cyclical measure. The individual suggested that if industry supported them, Ph.D.s could teach in secondary schools during periods of downturn in the industry cycle and provide a resource that can be rehired when the industrial cycle turns up again.

In another variation on Ph.D. involvement in research, it is important to note that some Ph.D. teachers also engage in active research during the school year with their secondary school students. Such Ph.D.s often involve their students in community-based research, for example on local environmental issues.

We noted in our interviews with science Ph.D.s that many pursue other professional activities besides research—perhaps more often than research. For example, while some did maintain research interests during the summer, others were engaged in producing teacher workshops, adapting computer programs to the high school setting, writing textbooks, reviewing grant applications for NSF and other foundation programs, or consulting.

Other recommendations Ph.D.s made for improving the work environment for Ph.D.s teaching secondary school science and math were:

- providing better laboratory equipment;
- providing more infrastructure support (classrooms, facilities etc.);
- hiring science aides who could set up equipment and experiments for teachers;
- dealing with students who have discipline problems;
- providing for smaller class sizes;
- improving the quality of the teachers, especially those who "taught to the test" just to get pay raises; and
- having a mix of teachers with and without Ph.D.s in the schools, since each has strengths to bring to the classroom and they can complement each other.

Finally, there were suggestions for the professional development of Ph.D. science and mathematics teachers. One Ph.D. who used to be vice president of a major university and is now a math teacher expressed interest in workshops offered for Ph.D.s who are teaching at the high school level. These workshops would concentrate on the substantive material and more on networking and discussion of ways to approach the material and develop effective communication with students. A state school official said that a program to bring science Ph.D.s to secondary school education and to support them in that environment should not just develop them as teachers, but also provide them an opportunity to become leaders in secondary school science and mathematics education in their states.

4

Discussion

The NRC urged in *National Science Education Standards* and subsequent reports that science teachers should bring new approaches, such as inquiry-based learning, to their classrooms. At the same time, the nation is faced with hiring as many as 200,000 public secondary school science and mathematics teachers over the next decade. This study has examined whether Ph.D. scientists and mathematicians could be attracted to secondary school science and mathematics education.

We have found that more Ph.D.s are potentially interested in such positions than is reflected in their current K-12 employment level. We also found that certain kinds of programs and incentives could provide conditions under which many Ph.D.s would consider secondary school science and mathematics education careers. In this chapter, we summarize our findings and discuss the pros and cons of the various incentives that survey respondents found attractive as a way of providing guidance to all of those interested in this important issue.

FINDINGS

Interest in Secondary Teaching

Based on our survey results, there appears to be sufficient interest by a reasonable number of Ph.D.s in secondary school teaching for the NRC to continue to investigate whether to develop and how to implement a program to attract Ph.D.s to secondary school teaching positions. We do not know exactly what number or percentage of Ph.D.s might ultimately consider secondary school teaching as a career. As with any career choice, this would depend on the specific incentives offered and the alternatives available at the time of choice. However, our survey results do demonstrate that interest in careers in secondary school science and mathematics education appears to be much higher among Ph.D.s than is represented by the 0.8 percent of them who currently work in K-12 education. This interest is high enough, we believe, to justify demonstration programs to test the feasibility of this career alternative.

Respondents to our survey had considered at least four different options in contemplating their career futures. We found that 36 percent of respondents had considered secondary school teaching or other secondary education positions in their career decision making. We found that 10 percent had previously taught in an elementary or secondary school. This pool could provide an untapped source for helping to meet the needs for secondary school science and mathematics education.

Among those we surveyed, individuals with certain characteristics were more likely than others to have considered a position in secondary school science or mathematics education. Respondents who were still in graduate school or who were not currently in postdoctoral positions were the most open to considering secondary school education positions. Graduate students and recent Ph.D.s in the biological sciences were more likely to consider secondary school teaching positions. Physicists and mathematicians were the most likely. Also women were more likely than men, and U.S. citizens were much more likely than non-U.S. citizens, to have considered or to consider secondary school positions. We found that chemists were least likely to consider such positions. Presumably they have strong opportunities in industry as well as academe.

Although some focus group participants held negative perceptions about secondary school teaching, many held a number of positive perceptions. These included attractive working hours, a work schedule the same as that of their children, and time for research or other activities during the summer. More to the point, many focus group participants also believed they would enjoy fostering scientific interest among students.

We also found through our survey that many recent science and mathematics Ph.D.s would consider other secondary school education positions in professional development, in university- or industry-based science education partnerships, or as science or mathematics specialists in school districts or a regional science resource center. Indeed, the Ph.D. teachers we interviewed indicated that, in addition to teaching, they currently provide leadership through curriculum development, professional workshops for teachers, and writing new textbooks.

Challenges

Given our survey results, a key question that a project designed to attract Ph.D.s to secondary school teaching must address is why, if up to 36 percent of science and mathematics Ph.D.s have considered secondary teaching careers, are less than 1 percent currently employed by K-12 educational institutions?

Some of the answers to this question are that Ph.D.s have many negative perceptions about secondary school education that prevent them from considering secondary school teaching positions. Some participants in our focus groups perceived a potential lack of status and respect as teachers, poor classroom laboratory facilities, excessive numbers of students in classrooms, structured curricula that allowed little opportunity for creativity, possible conflicts with non-Ph.D. teachers, and discipline problems. Both focus group discussions and survey results indicated that graduate students and recent Ph.D.s often perceive little added value in taking a full complement of education courses, and teacher certification is perceived as a barrier that is difficult to overcome.

Salary expectations also present a challenge for attracting Ph.D.s to secondary school education. Our survey found that respondents who had considered a career in secondary school education had lower salary expectations for a secondary school teaching position than those who had not, both for starting salaries and salaries anticipated in 7 years. All survey respondents recognized that salaries for positions in secondary school teaching were lower than for other career options. However, the average starting salary for teachers anticipated by graduate students and Ph.D.s in our survey—$37,400—is at the high end of the range of starting salaries offered to Ph.D.s by school districts nationwide.

There are many stereotypes about Ph.D.s that create additional challenges. Many school administrators argue that Ph.D.s may have good content knowledge, but do not have necessary pedagogical skills or cannot relate to secondary school students. Similarly, many university faculty do not encourage nonacademic careers for Ph.D.s generally, much less careers in secondary school education. Indeed, graduate students typically aspire to positions in academic science and mathematics similar to those of their mentors, and the socialization process in graduate school reinforces this. In some cases, graduate students who have expressed an interest in secondary school education careers have been treated as second-class citizens. As a consequence, many graduate students may fear that they will compromise their careers by vocalizing an interest in secondary school education.

What is Necessary for Success?

Given these challenges, what is necessary for success?

A program to attract Ph.D.s to secondary school teaching must do what it can to overcome negative perceptions about working in the secondary school environment. To combat such perceptions as low status and lack of respect, excessive numbers of students in classrooms, possible conflicts with non-Ph.D. teachers, student discipline problems, and lower salaries, a program to attract Ph.D.s to secondary education should focus on recruiting Ph.D.s for whom the perceived positives outweigh the negatives.

It is clear that Ph.D.s who seek secondary school careers must be properly motivated in order to be successful. Those who select a secondary school teaching career also need to assess their own personalities and interests. Those who will succeed and find fulfillment in secondary school education will be those who love teaching and enjoy helping students learn and achieve. Ph.D. teachers we interviewed told us that, for them, the love of teaching and the enjoyment they get from working with children helps compensate for lower salaries than Ph.D.s could command in other science and mathematics careers.

There were differences of preference among focus group participants about whether they would like to teach at regular public high schools or science and technology magnet schools. Many thought they would enjoy teaching in magnet or private schools. One participant said that "teaching at a good secondary school could be better than teaching at a bad college." However, other participants thought that teaching in regular public secondary schools where they could motivate regular students would be more socially responsible. The survey results also indicated that teaching in a magnet school would be a strong attraction for Ph.D.s. These differences in preference as well as differences in needs among school districts suggest that flexibility in placement of Ph.D.s with schools would provide the most appropriate course.

Results from both the focus group and survey indicated that graduate students and recent Ph.D.s often perceive little value in education courses and see teacher certification as a barrier to pursuing a teaching careers. However, these barriers could potentially be overcome by designing a program of courses in educational pedagogy tailored to the needs of Ph.D.s.

Both the Ph.D. teachers and school administrators we interviewed strongly recommended that Ph.D.s take education courses and obtain certification to teach. Ph.D. teachers believe that there is value in education courses and that certification is an important outward sign of professional acceptance. The consensus among Ph.D.s we talked to and those we surveyed, however, is that accelerated programs leading to certification that are tailored to their needs would be the most effective way of attracting and preparing them for secondary school careers. In addition, Ph.D.s who seek secondary school careers should also have relevant experience and education. Ph.D. teachers and administrators strongly recommended that those considering a secondary school teaching career should seek opportunities to work in classrooms, both to obtain direct teaching experience and to explore whether teaching career is right for them.

Our interviews with Ph.D. teachers and school administrators also indicated that, while negative stereotypes about Ph.D.s as secondary school teachers are widespread, they have not posed obstacles to all Ph.D.s who seek secondary school careers. For example, we were pleased to learn that, in practice, only a minority of the Ph.D. teachers we spoke with had encountered resistance from school administrators or teachers based negative stereotypes, but this is a widespread conviction. Also, we found in talking to Ph.D.s teachers that only a minority faced active disregard from colleagues and mentors after announcing a decision to take a secondary school position.

Ph.D.s could be attracted to secondary school education through programs that address their needs and interests and sustain them as teachers. We presented graduate students and recent Ph.D.s who participated in our survey with a number of scenarios and asked whether they would consider secondary school science or mathematics teaching under each of them. Respondents indicated strong interest in teaching if they were awarded a prestigious national fellowship that would provide training, placement, and special opportunities for networking with peers, as well as cover living expenses during the training period. They also gave strong support to consideration of secondary school teaching if there were an accelerated certification process, mentoring in curriculum development and pedagogy, resources that support science education, and higher salaries.

Participants in our focus groups also indicated that poor classroom laboratory facilities were a disincentive to taking secondary school teaching positions. There may, however, be creative ways that states and school districts can compensate for this. For example, respondents to our survey were attracted to the idea of having teaching support from a regional- or university-based science teaching resource center that provided science kits, loaned laboratory equipment, and organized field opportunities for student science experiments in which students could participate.

We also found through our survey that many Ph.D.s seek the option of having continued involvement in science and, potentially, taking a leadership role in science and mathematics education. Having funded summer research opportunities and funded attendance at professional meetings during the school year is appealing to graduate students and recent Ph.D.s who have considered secondary school careers. Many Ph.D.s would also potentially seek opportunities to help build a bridge between K-12 and postsecondary science and mathematics and to take on other leadership roles in K-12 science and mathematics education other than classroom teaching.

Finally, we found that many survey respondents would only make an initial two-year commitment to secondary school education positions. However, this should not be seen as limiting any activity designed to attract Ph.D.s to such positions. We believe it is important to provide Ph.D.s a means for obtaining such positions, allowing them to realize their own potential long-term career interests, and provide them incentives to stay longer. Among those incentives might be special training opportunities that would facilitate career options in other K-12 leadership roles as well as teaching positions.

GUIDANCE FOR DEMONSTRATION PROGRAMS

As a next step, the NRC should continue to explore the development of possible demonstration programs. In developing such programs, the committee overseeing the second phase of the NRC's project on attracting Ph.D.s to secondary science and mathematics education should consider a number of potential programmatic features that could help to make secondary school teaching attractive to Ph.D.s. These features could include demonstration programs developed and administered by states, a national fellowship program designed both to attract and select appropriate candidates, a rigorous process for selecting Ph.D.s for secondary school teaching, and flexibility in placing project should also consider the role of postsecondary institutions in the demonstration programs and the implementation of an evaluation as a critical ongoing component of the demonstration.

Incentives for attracting Ph.D.s to secondary school teaching that the committee overseeing the second phase of the project should consider include an intensive summer program in education and pedagogy, funding for such training, and an accelerated certification process. The availability of mentors for at least the first year of teaching, additional teaching resources, salary supplements, opportunities to network with peers and scientific colleagues, and summer research opportunities are additional features that should be considered.

Organization of Demonstration Programs

State Demonstration Programs

The committee overseeing the second phase of this project should develop a design for demonstration programs that could be carried out in cooperation with a small number of interested states. We recommend that state governments organize these demonstrations because states play a stronger role than the federal government in education in the United States and they can potentially bring more resources to bear than can local school districts. States would also develop their demonstration programs to fit their own educational and human resources needs.

We expect that state programs will differ from one another in their features. An evaluation of the demonstration programs would take the diversity of state approaches into account in testing the feasibility and effectiveness of attracting Ph.D.s to secondary school classrooms.

Selection of Ph.D.s for Teaching Positions

The committee recognized that the identification and selection of science and mathematics Ph.D.s for secondary school teaching is a critical step in the process of implementing demonstrations. The selection process should identify individuals who have strong knowledge of their subject matter, a demonstrated interest in secondary school science and mathematics education, and personal characteristics appropriate to the secondary school education environment.

Role of Postsecondary Institutions

The phase two committee should consider designing desmonstration programs that have strong linkages to science and mathematics programs at colleges and universities in their states, piggybacking on any existing partnership programs. Colleges and universities could facilitate the recruitment and preparation of Ph.D.s for secondary school teaching and provide opportunities for classroom and secondary school experience for Ph.D.s interested in applying to the demonstration programs. They could also serve as venues for special workshops and meetings for Ph.D. teachers during the school year as part of a demonstration program in a given state and provide resources to support secondary school science and mathematics education. These kinds of connections would provide opportunities for interchange between the K-12 educational system and postsecondary science and mathematics that would benefit both and would also create a community of individuals who can bridge the gap between K-12 and undergraduate approaches to science and mathematics education.

Flexibility in School Placement

The state demonstration programs ideally would place and support Ph.D. s in a variety of secondary school education positions, including teaching positions in regular public secondary schools and science and technology magnet schools, as appropriate to the needs of the state and the fellows. The ability to teach in a specialized science and technology high school was very popular among respondents. Fifty-six percent of all respondents and 81 percent of those who have considered secondary school careers indicated that they would more strongly favor secondary school teaching with such an option. We learned through our focus groups, however, that some Ph.D.s would only consider teaching in traditional, public secondary schools, because they wanted to be able to help students in these schools. Therefore, while we recognize that states may want to concentrate Ph.D.s in magnet schools, we do not believe that they should place them exclusively in such schools. They should approach placement flexibly so as to meet the needs and interests of both Ph.D.s and school districts.

Regional Clustering

The committee overseeing the second phase of this study should consider regional clustering of Ph..D.s in the state demonstration programs to facilitate networking, to optimize use of laboratory resources and science teaching resource centers, and to forge links between demonstration programs and university education and science departments.

Evaluation

Finally, demonstration programs should include an evaluation component to be implemented simultaneously with the demonstrations. Any evaluation should have clear and well-defined goals. The phase two committee should develop an evaluation plan for each demonstration project that addresses the goals for the third phase of the study. This phase three evaluation plan must address a wide range of goals that focus on the feasibility of placing science and mathematics Ph.D.s in secondary school teaching, assessing the process of implementing such a program, and conducting an outcome evaluation based on measurable goals. Such an evaluation should focus on whether the programs in fact demonstrate that attracting Ph.D.s to secondary school teaching is a feasible, cost-effective, and worthwhile activity. A cross-site evaluation of the state demonstration programs, including their means for recruiting, placing, and supporting Ph.D.s, would highlight strategies for attracting Ph.D.s to secondary school education that were effective or ineffective in each of the states. A final report on the state demonstrations would synthesize program outcomes to assist other states that might decide to develop similar programs.

Pedagogical Skills and Teaching Resources

Education Courses and Certification

Based on interview and survey results, we strongly support the development of education courses and a teacher certification process tailored to the experiences and needs of Ph.D. scientists and mathematicians. Our survey showed that Ph.D.s would clearly be interested in considering teaching positions if they could demonstrate teaching proficiency on the job rather than take courses in educational psychology and pedagogy first. However, interviews with administrators and Ph.D. teachers indicated that education courses do provide teachers with important pedagogical knowledge and that certification is an important step in establishing oneself as a teacher. We agree.

We strongly suggest that any program designed to attract Ph.D.s to secondary school teaching have a pedagogical component but, as a practical matter, it is not likely to be attractive to our target group of already highly educated individuals unless it can be accomplished in a fairly compressed manner. We found that 44 percent of our survey respondents and more than two-thirds of the respondents who had previously considered teaching careers indicated that they would consider a teaching position if they could receive teacher certification by taking an intensive summer course in education. The percent who would consider teaching if the period of time were increased to one year dropped precipitously to just 14 percent overall and 22 percent for those who had previously considered teaching. Thus, an extended period of coursework would be a strong disincentive to recruitment. Ph.D. teachers we interviewed—who strongly supported education courses and certification for Ph.D.s seeking to teach at the secondary level—indicated that they believed a streamlined course in educational theory and practice leading to certification could be developed for Ph.D.s.

State demonstration programs, in conjunction with a national fellows program, if created, might provide Ph.D.s with an intensive summer program in educational theory and practice, and fund participants during this summer program. The state demonstration programs should provide a process by which Ph.D.s could obtain teaching certification in an accelerated manner. The summer program should focus on educational psychology, pedagogy, and pedagogical content knowledge. Ph.D. scientists and mathematicians could bring strong knowledge of science and mathematics to the classroom, as well as experience in scientific inquiry. To fully translate this into the inquiry-based teaching and learning that we would like Ph.D.s to bring to the classroom, Ph.D.s must be trained to understand how students learn.

Mentoring

The committee suggests that states should select and appoint experienced teachers to serve as mentors to Ph.D.s participating in their demonstration programs. Our survey clearly indicated that Ph.D.s would be more likely to consider teaching positions if they had the support of a mentor. Almost half of our respondents, and more than two-thirds of those who had previously considered teaching indicated that they would consider such positions if they had a selected, experienced teacher to mentor them during for their first year of teaching or on an as-needed basis. Providing mentors will potentially add programmatic costs if states provide additional compensation to mentors, but the availability of mentors would be a programmatic feature that could help induce Ph.D.s to teaching careers.

Science Teaching Resources

One of the negative perceptions that participants in our focus groups held about secondary school teaching is that they would be working with inadequate laboratory equipment. Through our survey, however, we found that 52 percent of respondents and three-quarters of respondents who had previously considered teaching, would consider a secondary school teaching career if they received support from a regional- or university-based science resource center that provided science kits and loaned equipment.

The committee overseeing the second phase of this project should work with states and school districts to determine whether the development of such science teaching resource centers would be a feasible component of state demonstration programs. We believe that they would be useful as one means for attracting Ph.D.s to secondary school teaching and also for improving the quality of science and mathematics education generally.

Future Positions for Ph.D.s

We strongly believe that Ph.D.s can contribute not only in the classroom, but also in other K-12 science and mathematics education positions. The most popular positions among our survey respondents were professional development (teaching science or mathematics teachers), science or mathematics specialist for a school district, working in a university- or industry-based science educational partnership, or serving as a science specialist in a science resource center. To a lesser degree, there was also interest in curriculum development or work in a science museum, environmental science center, or similar institution. While Ph.D.s could eventually contribute as leaders in K-12 science and mathematics education through these positions, it would be useful for many of these positions, if not all, for Ph.D.s to have secondary school teaching experience first. It is important to have classroom experience before taking on broader, leadership roles.

Incentives

The survey identified a number of incentives that respondents indicated would favorably affect their consideration of assuming a position in secondary school teaching. A description of these incentives is presented in the following pages.

National Fellowship Program

National Fellowship Program. Through our survey we found that two-thirds of our respondents and almost 90 percent of respondents who had previously considered secondary school careers would consider taking a position as a secondary school teacher if they were awarded a fellowship that provided training, placement, and special opportunities for networking with peers, and covered living expenses during the training period. Given the potential such a fellowship might have for attracting Ph.D.s to secondary school teaching, the phase two committee should consider ways in which a fellowship program might be established and administered by a prestigious national agency or organization. The national program, instituted in cooperation with the states, could select and train fellows, fund them during their training, and provide an ongoing opportunity for networking with peers.

A program that is national in scope could potentially attract resources from national sources, draw applicants from across the country, and serve as a catalyst for the state demonstration programs. A prestigious fellowship program would attract applicants who might not otherwise consider secondary school positions and produce a cohort of science and mathematics teachers who could conceivably change the way science and mathematics are taught. The potential downside to a "prestigious" fellowship for Ph.D. teachers is that it might adversely differentiate them from the population of teachers we want them to join. We also recognize that the establishment of a national fellowship program would increase costs.

Compensation. Survey respondents recognized that salaries for secondary school teaching were lower than for other career options. Still, the average starting salary for teachers anticipated by graduate students and Ph.D.s in our survey—$37,400—is within the range of starting salaries offered to Ph.D. teachers by school districts, albeit at the high end of the range. We believe that states and school districts will need to demonstrate a strong financial commitment to the program for it to succeed by supplementing Ph.D. salaries. For example, by providing stipends for attendance at scientific meetings, and for other activities related to the professional development of Ph.D.s and the benefit of their students. Some states will be in a better position than others to offer these financial incentives and thus to attract Ph.D.s to teaching positions. The committee does not have evidence to assess whether such adjustments by states with lower starting salaries can realistically be expected.

Peer Networking

Our survey appeared to demonstrate that, as teachers, Ph.D.s would welcome the opportunity to continue to network with their professional peers. Those surveyed responded very positively to consideration of teaching if they were awarded a prestigious national fellowship that provided special opportunities for networking with peers. If a national fellowship program is developed, such a program could include an annual meeting of fellows. Whether or not there is a national fellows program, states should consider regional clustering of Ph.D.s to facilitate networking opportunities. Opportunities for Ph.D.s in secondary school education to network with others in similar positions would allow them to share ideas in curriculum development, how to present material in classrooms, and other secondary school education issues.

Connections with the Larger Scientific Community

State demonstration programs should provide opportunities for interactions between Ph.D. teachers and the scientific community in academia and industry.

We found that 63 percent of survey respondents and 88 percent of respondents who had previously considered teaching careers would consider teaching if they were given funding and time to attend two or more scientific meetings during the school year. Even if they were given the opportunity to attend only one meeting during the school year, there was still considerable interest in teaching if this was an option. Thus, states should consider facilitating attendance at professional meetings as one feature of demonstration programs.

We also found that 63 percent of respondents and 79 percent of those who had previously considered teaching would now consider teaching if they were guaranteed a summer fellowship, with travel expenses, in a research laboratory. States should consider how they could develop links to universities and businesses to provide summer research opportunities for Ph.D.s, as some already do for science teachers. This would provide an incentive for Ph.D.s to take teaching positions, but it would also provide a means for allowing Ph.D. teachers to stay involved in scientific research and connected to developments in scientific knowledge and techniques. This ability to stay current would contribute to their ability to provide quality science and mathematics education to their students.

The notion that some Ph.D.s might be attracted to secondary school teaching if they could continue to participate in research over the summer is not designed to signal that research is still more favored than teaching. Rather it is designed to find a way to make teaching and research complementary parts of a career rather than a dichotomous choice. Furthermore, it would be valuable for at least some secondary school science and mathematics teachers to remain connected to research so that current methods and knowledge are circulated throughout the educational system.

REFERENCES

Association of American Universities. 1998. "Committee on Postdoctoral Education: Report and Recommendations." Washington, DC: Association of American Universities.

Boyer Commission on Educating Undergraduates in the Research University. 1998. *Reinventing Undergraduate Education: A Blueprint for America's Research Universities.* Menlo Park, CA: Carnegie Foundation for the Advancement of Teaching.

Commission on Professionals in Science and Technology. 1997. "Postdocs and Career Prospects: A Status Report." Washington, DC: Commission on Professionals in Science and Technology.

Feistritzer, C.E. and Chester, D.T. 1998. *Alternative Teacher Certification: A State-by-State Analysis 1998-99.* Washington, DC: National Center for Education Information.

Gitomer, D.H, Latham, A.S., and Ziomek, R. 1999. *The Academic Quality of Prospective Teachers: The Impact of Admissions and Licensure Testing.* Princeton, NJ: Educational Testing Service and ACT, Inc.

Ingersoll, R. 1999. "The Problem of Underqualified Teachers in American Secondary Schools." *Educational Researcher, 28,* pp. 26-36.

Kennedy, D. 1997. *Academic Duty.* Cambridge, MA: Harvard University Press.

National Academy of Sciences. 1995. *Reshaping the Graduate Education of Scientists and Engineers.* Washington, DC: National Academy Press.

National Center for Education Statistics. 1999a. *Projections of Education Statistics to 2009.* (NCES 1999038) By Debra E. Gerald and William J. Hussar. Washington, DC: National Center for Education Statistics.

National Center for Education Statistics. 1999b. *Projecting the Need for Newly Hired Teachers in the United States.* (NCES 1999026) By William J. Hussar. Washington, DC: National Center for Education Statistics.

National Education Association. 1997. *Status of the American Public School Teacher, 1995-1996.* Washington, DC: National Education Association.

National Research Council. 1996. *National Science Education Standards.* Washington, DC: National Academy Press.

National Research Council. 1997. *Science Teaching Reconsidered: A Handbook.* Washington, DC: National Academy Press.
National Research Council. 1998. *Trends in the Early Careers of Life Scientists.* Washington, DC: National Academy Press.
National Research Council. 1999. *Global Perspectives for Local Action: Using TIMSS to Improve U.S. mathematics and Science Education.* Washington, DC: National Academy Press.
National Science Foundation. 1996. *Indicators of Science and Mathematics Education, 1995.* Arlington, VA: National Science Foundation.
National Science Foundation. 1999. *National Graduate Teaching Fellows in K-12 Education: Program Announcement.* (NSF 99-75) Arlington, VA: National Science Foundation.
Rice, R.E. 1996. "Making a place for the new American scholar." Inquiry #1. *Working Paper Series on New Pathways: Faculty Careers and Employment for the 21st Century.* Washington, DC: American Association for Higher Education.
Seymour, E., and Hewitt, N. 1997. *Talking About Leaving: Why Undergraduates Leave the Sciences.* Westview Press.
Tobias, S., Chubin, D.E., and Aylesworth, K. 1995. *Rethinking Science as a Career: Perceptions and Realities in the Physical Sciences.* Tucson, AZ: Research Corporation.

Appendix A

Study Methodology

INTRODUCTION

In carrying out this study, the committee—with staff from the NRC's Office of Scientific and Engineering Personnel (OSEP) and Center for Education —conducted a national survey and a series of telephone interviews.

To obtain data on the willingness of Ph.D.s to consider positions in K-12 science and mathematics education and on the various conditions that might enhance their recruitment, OSEP staff conducted a national survey of graduate students and postdoctorates. To inform development of the survey questionnaire, staff conducted five focus groups on career aspirations and perceptions of K-12 science and mathematics education careers with advanced graduate students and recent Ph.D.s. Led by a professional facilitator, focus groups obtained reactions to hypothetical incentives to attract Ph.D.s to secondary school positions and also elicited additional ideas that were incorporated into the survey questionnaire.

Focus group participants represented the spectrum of fields in the biological sciences, physical sciences, and mathematics. The focus groups were also located geographically diversified, with sessions held at Rutgers University/University of Medicine and Dentistry of New Jersey, Duke University, the University of Texas at Austin, the University of California, San Francisco, and the Fred Hutchison Cancer Research Center in Seattle.

The survey questionnaire reflected the input from these focus groups as well as input from committee members and staff from several NRC units. It assessed four broad areas of interest: (1) demographic characteristics of the respondents; (2) their short-term and long-term career aspirations, including salary expectations; (3) conditions under which the respondent might consider K-12 science and mathematics education as a career; and (4) incentives that might be required for respondents to consider careers in K-12 science and mathematics education.

In addition to this survey, staff conducted a series of telephone interviews with interested individuals in K-12 science and mathematics education during summer 1999. Interviews were conducted with 18 science and mathematics Ph.D.s currently teaching in secondary schools to synthesize their experiences, including barriers encountered, strategies adopted to overcome these barriers, areas of success and achievement, and tips on recruitment and retention of other science Ph.D.s. Interviews were also conducted with high school principals, school district superintendents, and chief state school officers, to obtain information on concerns with employment of Ph.D.s in secondary school, certification issues, and funding of Ph.D.s in secondary education. In addition, telephone interviews were conducted with graduate deans to ascertain the kinds of programmatic changes required in graduate education to prepare Ph.D.s for careers teaching in secondary schools.

SURVEY OF GRADUATE STUDENTS AND RECENT PH.D.S

The survey was fielded in July and August 1999 to a national sample of 2,000 graduate students and recent Ph.D.s stratified by field of study. This survey gathered data on demographic characteristics, short-term and long-term career aspirations, salary expectations, and conditions under which respondents would consider employment in secondary school education. The NRC received 713 responses to the survey. A copy of the survey instrument is located in Appendix C.

Focus Groups

As a first step in survey development, five focus groups were conducted to gain greater insight to the attitudes of science and mathematics graduate students and postdocs on secondary school education activities. The focus groups were conducted in sites that included both life scientists and physical scientists/ mathematicians, graduate students and postdocs, and represented a broad geographical spread. The focus group protocol focused on six broad areas, including:

- Career aspirations.
- Consideration of aspects of careers other than in a university.
- Considerations of careers in secondary education.
- Perceptions of teacher certification.
- Perceptions of teacher compensation.
- Perceptions of work the environment in secondary education.

A copy of the focus group protocol is located in Appendix B. An integrated summary of the five focus groups is located in Appendix C.

Questionnaire Development

Based on input from the focus groups, a questionnaire was designed to ascertain conditions under which graduate students and recent postdocs in science and mathematics would consider employment in secondary school education. In addition to questions on demographic and educational characteristics, the questionnaire proposed a number of scenarios under which the respondents would consider secondary school teaching. Questions were addressed to other aspects of secondary school employment such as curriculum and professional development.

The questionnaire was pretested with a small group of graduate students and recent Ph.D.s in Washington, D.C. and revised again. A copy of the questionnaire is located in Appendix C.

Sampling

The questionnaire was then sent to a random sample of 2,000 science and mathematics graduate students and postdocs. NRC staff developed a stratified random sample of science and mathematics graduate students and postdocs for this survey. The sampling frame was derived from the 1993 edition of *Research-Doctorate Programs in the United States: Continuity and Change.* The population of sampling units consisted of science and mathematics departments listed in *Research-Doctorate Programs*—a total of 935 life science and 570 physical science/mathematics departments. The committee had concluded that astrophysics/astronomy, computer science, and statistics/biostatistics departments were not germane to this survey and these departments were excluded from the sampling frame. The sample was stratified into life science and physical science/mathematics categories using proportional allocation with proportionality based on the number of graduate students in physical science/mathematics and life science departments as listed in the 1993 *Research-Doctorate Programs* appendixes.

A two-stage sampling frame (with replacement) was developed from the listing of Ph.D. granting departments in *Research-Doctorate Programs in the United States.* In the initial stage, departments were selected using systematic sampling, stratified by discipline. Systematic sampling selects items from an ordered array by selecting a random starting point and selecting items on periodic basis. For example, for the life sciences, we identified a random starting point between 1 and 39 and then selected that and every 39th item in the array. Sampling with replacement provides an exact replacement sample item for each nonresponding department by identifying a specific replacement unit for each sampling unit. This was accomplished by identifying pairs of departments in the systematic sample. The first department was designated as the primary sampling unit, and the nest department on the listing was identified as the replacement-sampling unit.

Field of Study	Number of Departments
Life Sciences	
Biochemistry and Molecular Biology	187
Cell and Developmental Biology	165
Ecology, Evolution, and Behavior	127
Molecular and General Genetics	102
Neuroscience	98
Pharmacology	121
Physiology	135
Natural Sciences and Mathematics	
Chemistry	168
Geosciences	95
Mathematics	135
Oceanography	26
Physics	146

According to *Research-Doctorate Programs,* there were an estimated 30,500 life science and 37,594 natural science and mathematics graduate students in the identified departments in 1993. Life science departments had an average of 32.6 graduate students while physical science/mathematics departments had an average of 65.9 graduate students. The sampling frame consisted of 47 departments—17 physical science/mathematics and 30 life science departments. Approximately one-fourth of the departments is ranked in each quartile of scholarly quality. However, because higher ranked departments have more graduate students, there was a disproportionately larger number graduate students in the sample from higher ranked departments. For example, among chemistry departments top-ranked departments had an average of 180 students. For chemistry departments ranked in the second, third, and fourth quartiles, the average number of graduate students were 88, 53, and 31, respectively. Approximately 42 percent of graduate students/postdocs in the sample were attending universities ranked in the top quartile. For the second, third, and fourth quartiles, the percentages were 23, 27, and 8 percent, respectively. At the second stage of sampling, all doctoral graduate students and postdocs at a selected department are included in the sample and received a questionnaire.

The chairs of the 47 departments were contacted by a letter that explained the nature of the survey and requested a listing of the names and addresses of all graduate students and postdocs. A total of 38 departments, from either the primary or nonrespondent lists, submitted rosters in time to be included in the survey, resulting in 2,713 graduate students and postdocs identified in the first stage of sampling. Approximately 55 percent of these were in life sciences. Using systematic sampling, questionnaires were sent to 2,000 of the graduate students and postdocs. Depending on the locator data provided by the department, graduate students and postdocs were contacted by either mail (1,144) or email (856). Two departments were unwilling to provide individual student locator data, but agreed to submit questionnaires sent in bulk through campus mail. Consequently, 241 of the graduate students/postdocs contacted by mail had the packet of questionnaire, cover letter, and return envelope distributed through campus mail rather than by direct mail.

Survey Response

A total of 2,000 science and mathematics graduate students and postdocs were surveyed in three waves on July 15, August 9, and September 2 of 1999. A total of 715 responses were received; 680 responses were received before the cut-off date for data entry. For 169 subjects, no delivery could be made by either mail or email. An additional, nine subjects were out-of-the scope of the survey. The response rate was 39 percent.

There are two possible reasons for the low response rate. First, if subjects were biased against secondary school education activities, they might refuse to respond to the survey. Second, nearly all of the subjects were contacted through their university addresses (by either mail or email) and if these respondents were not located at their university locations during the summer, this could account for the low response rate.

To test whether there was a nonresponse bias, NRC staff conducted a telephone survey with a random sample of nonrespondents. Nonrespondent candidates were selected using systematic sampling from a listing of nonrespondents, a method of selection designed to ensure that the sample was representative of all nonrespondents. A sample of 200 nonrespondents was selected and telephone numbers were obtained for 174 of them.

The focus of the survey was to determine why the subject had not responded to the survey. The telephone survey results are shown below.

Response	Number
Already responded	22
Conducted survey on telephone	1
Never received instrument	10
Received, but misplaced – send new survey	14
Too busy to respond	20
Out of the office – did not respond	12
Does not apply	6
Not interested in responding	3
Other response	4
Colleague answered – subject on vacation	21
Left voice mail message	36
No answer	25
Total	174

We believe that the low level of response is result of the timing of the mailout rather that any systematic bias of subjects toward the survey and the survey's content. During two days of calling, NRC staff were able to make contact with 113 persons. Voice mail messages, urging completion of the questionnaire, were left for 36 persons. There was no answer to repeated calls for 25 persons. Among those with whom NRC staff spoke, nearly half indicated that they had not responded to the questionnaire because they had not been at the office or had been too busy to respond. About 20 percent indicated that they had already responded (one completed the survey over the telephone). Only three persons indicated that they were not interested in K-12 education and had not responded to the survey for that reason. The largest category, 61 persons, represented those with whom NRC staff were unable to establish contact, even after repeated calls. Eventually NRC staff left voice mail messages when possible. However, there was never any contact with 25 persons.

On the basis of the telephone survey, we concluded that the low response rate was probably a consequence of survey administration during the summer when many graduate students are not at their office locations. The telephone survey found that practically none of the persons contacted refused to participate because of a bias towards secondary school teaching.

INTERVIEWS

In addition to fielding the national survey, the committee also collected information for the study through a series of interviews with Ph.D.s already in K-12 education careers and with administrators in K-12 and higher education.

Interviews with Ph.D.s

To complement the survey of graduate students and recent Ph.D.s about their potential interest in K-12 education, we interviewed 18 Ph.D.s already working at the K-12 level to draw insights from their experiences about the opportunities for and obstacles to Ph.D.s taking positions in K-12 education. A copy of the interview protocol is located in Appendix C.

These interviews were conducted by telephone during August 1999. NRC staff compiled a short list of Ph.D.s working in teaching or curriculum development positions and these individuals were contacted for an interview. At the end of each interview, we asked the individual being interviewed if he or she could think of someone else who would be suitable for interviewing. This "snowball" sampling method yielded more than 32 names, of which 20 were available for interviews in the time allotted. Two of the 20 did not hold doctorates, so the total number of Ph.D.s interviewed was 18.

This was not a random sample of persons with science and mathematics Ph.D.s who hold positions in K-12 education. The goal of these telephone interviews was not to obtain scientifically accurate survey data. The goal of these interviews was to provide background for and insights to the data collected through *The Survey of Graduate Students and Recent Ph.D.s.*

In interviews typically lasting 35 to 50 minutes, we asked these individuals to talk about their experiences in K-12 education. We asked them to comment on what attracted them to secondary school teaching or to curriculum development. We asked them whether colleagues or mentors were supportive when they decided to enter K-12 education, to describe any barriers they encountered in taking K-12 education positions, and how they handled teacher certification. We also asked for the effects of financial or family considerations on their career paths. Individuals interviewed were asked to comment on their current work environment, their relationships with other faculty and administrators, and whether they were actively engaged in research. Finally, we asked them whether they were happy with their choice of a career working in K-12 education, what advice they would give a Ph.D. considering such a career, and what the prospects for such a career would be at this time.

While the interviewer asked these specific questions, the interview was also conducted in such a manner to encourage the interviewee to explore the reasons why they went into teaching and to discuss what were the most significant issues involving that career choice.

Interviews with Administrators

In addition, telephone interviews were conducted with a number of state and local school administrators and higher education administrators, including:

- Chief state school officers.
- School district superintendents.
- High school principals.
- Principals of science and technology magnet schools.
- Graduate deans.

Interviews were conducted with high school principals, school district superintendents, and chief state school officers, to obtain information on concerns with employment of Ph.D.s in secondary school, certification issues, and funding of Ph.D.s in secondary education. In addition, telephone interviews were conducted with graduate deans to ascertain the kinds of programmatic changes required in graduate education to prepare Ph.D.s for careers teaching in secondary schools.

A professional skilled in in-depth interviewing conducted all of these interviews. A semi-structured interview schedule was developed that paralleled the interview guide for Ph.D. teachers. A copy of the interview schedule is located in Appendix B.

APPENDIX B

Focus Group Summaries

SITES OF FOCUS GROUPS

University of Medicine and Dentistry of New Jersey - Rutgers University/UMDNJ
Duke University
University of Texas at Austin
University of California, San Francisco (UCSF)
Fred Hutchinson Cancer Research Center (FHCRC)

SUMMARY OF FOCUS GROUPS

Interviewees - A total of 43 Ph.D. graduate students and postdoctoral fellows participated in the five focus groups. Most were in the biological sciences. A few were in mathematics and chemistry. One was in geological sciences. Several had started out to become teachers or had taught high school. Some had participated in secondary school teaching through the Science and Health Partnership (SEP).

Career Aspirations - Many would stay in scientific research if they could. Attitudes about possibility of obtaining tenure-track research positions were realistic. About one-third aspired to work in small colleges, another third in industry. Some still hoped for tenure track research positions. Perceptions of the academic professor's life were often negative. Several interviewees had considered teaching as a career. Almost none were planning to teach K-12.

Positive and negative attitudes about teaching K-12 - Although the importance and social responsibility of good science teaching was recognized and many had enjoyed teaching as teaching assistants, few considered K-12 teaching as an option. The perceived positives included great hours, schedule is same as your children, great vacations, summer option to do research, and enjoyment of fostering scientific interest among children. Some believed student motivation to learn at private and magnet schools would be a good. As one interviewee said, "Teaching at a good secondary school could be better than a bad college." Others believed this would be a socially irresponsible copout. The perceived negatives included lack of status and respect, poor classroom laboratory facilities, excessive numbers of students in classroom, structured curriculum with no opportunity for creativity, possible conflicts with non-Ph.D. teachers, and problems of discipline, violence, or drugs. A Ph.D. is overeducated for K-12, said one participant. "I enjoy teaching math too much to be teaching 10th grade algebra."

Curriculum Development and Professional Development–While not previously considered an option by any of the participants, the idea of being involved in curriculum development and teaching science to teachers aroused substantial interest at all sites. Participants thought curriculum development was appealing because of the opportunity it would provide to look at the "big picture" of how all the pieces of science or math education fit together. They thought training science teachers would provide them a role in keeping teachers current with regard to the latest scientific advances.

Training and Certification - Participants realized that in addition to knowing their subjects, teachers need to learn how to teach and to deal with children. Standard teacher education courses received little enthusiasm. At several sites a mentored teaching experience was suggested to be important. There was some interest in a time-limited teaching experience of 1-2 years. Five years was considered too long. It was suggested that this experience should be mentored, and that it could be rewarded by a guaranteed NIH-funded fellowship. Also heard was: "If you take 2 years off to teach you won't get back on the research track."

Compensation - Career and life style issues were considered more important than salary. K-12 salaries were considered unattractive in contrast to other professional options.

Comments and Suggestions

"The best time to capture Ph.D.s for secondary school teaching is right after graduation."

"If you put an ad in Science or Nature with an established program to recruit Ph.D. scientists to secondary school teaching it would work. Lots of students would sign up."

Focus Group Questions and Responses

Tell me about your current position?

- Rutgers/UMDNJ: five postdocs; five grad students; all in life sciences
- Duke: one postdoc; one Ph.D. just graduated; two Ph.D. candidates; one MS/MAT
- UT Austin: one in geology, three in math, one chemistry, one microbiology, one biochemistry, and one biology
- UCSF: All 10 are life science graduate students. Six are in biochemistry, and one each in cell biology, biophysics, pharmaceutical chemistry, and immunology
- FHCRC: All are postdocs at the Fred Hutchinson Cancer Research Center. Three are working in gene regulation, three are working in mouse genetics, and one each is working in drug design, cellular growth regulation, transcription regulation, and molecular genetics. The range from first to sixth-year postdocs. Seven of the 10 had teaching assistant (TA) experience while in graduate school.

What are your career aspirations?

Rutgers/UMDNJ:

- Academic research/tenure-track faculty: five (one said she might consider secondary school or college teaching in her native Puerto Rico)
- Not sure: maybe industry: one
- Own business (contract research): one
- Don't know: two (one postdoc facing a difficult job market; one 2nd year grad student who hasn't faced the issue)
- One participant indicated that they all wanted to do research. That's why they are in grad programs or postdocs positions.

Duke:

- Academic research/tenure-track faculty: two (one explicitly said research without teaching)
- Undergraduate teaching at a liberal arts college: one
- Non-formal environmental education: one
- K-12 mathematics curriculum development, but career is on hold while husband finishes mathematics Ph.D.

UT Austin:

- Two were interested in teaching in a small college
- Four were interested in research, one was emphatic
- One wanted to work for the FBI as an analytical chemist (and not in academia)
- One wanted to work in industry
- All participants had teaching assistantships (university requirement) and, consequently, some teaching experience

UCSF:

- Only three out of 10 want research careers, several want to teach in small colleges. For the immediate future four have lined up postdocs, two want to teach (one at a community college and one at a secondary school), two are interested in industry, and two are interested in a traditional academic career. One of the students leaning toward academics is also considering patent law and three of the others have considered teaching at the community college level. All 10 are involved in the SEP program that gets them involved in high schools, but not at the teaching level. In addition, all students are required to serve as a TA at some time in their student careers. This is interesting as there are no undergraduate students, so they have 2-3rd year graduate students teaching 1st year students.
- One sees no practical applications of research and wants to get into industry
- One wants to teach at a small college, enjoys teaching and will give up cutting-edge research for teaching, wants to get out of the pressure cooker. This is definitely a life-style exchange.
- Another needs the big equipment that is only available at a big research university; likes cutting-edge research.
- Another commented that big universities have the big salaries but a lot of stress. Small colleges have lower salaries and lower stress. This is a trade off; teaching is not lucrative.

FHCRC:

- Two want to teach in small colleges, three want to remain in research (one has an interest in technical writing), two want to work in industry, one wants a traditional academic/research position (although has considered teaching), one is interested in science education administration, and one is undecided.

What do you find attractive about the career you aspire to?

***Rutgers/UMDNJ*:**

- I like science,
- Enjoy scientific investigation,
- Intellectual seeking,
- Independence, and
- Can develop your own experiments.

Duke:

- While not all of the participants explicitly answered this question, it was evident from the discussion of career aspirations that each was motivated by what they would find fulfilling. As with the Rutgers/UMDNJ group, they were focussed on the substance of what they would be accomplishing in their careers rather than considerations such as salary (see below).
- K-12 math curriculum: would enjoy reforming the math curriculum at a school.
- Environmental education: enjoys taking science teachers through experiments that they can teach to students.
- Neurobiological research: would enjoy applying skills in research.
- Undergraduate education: has taught secondary school English in Japan and enjoys teaching. Would feel overqualified for secondary school teaching; would enjoy undergraduate teaching.

***UT Austin*:**

- Teaching gives you a better return on your investment, the research process is too long for so little research to happen.
- Started in education but was bored, switched to math – research is a stressor, the roller coaster of research is too much.
- Started in elementary education but transferred into chemistry, too much lesson planning in elementary education.
- Started in chemical engineering but switched to biochemistry, didn't like the responsibilities associated with chemical engineering.

What is unattractive about the career you aspire to?

***Rutgers/UMDNJ*:**

- Having to pursue research grants.
- Salaries are low.
- The personal costs of obtaining the tenure-track faculty research positions are too high.

Duke:

- See "chance of realizing aspiration" below.

***UT Austin*:**

- Too many Ph.D.s, competition is tough, computer science and electrical engineering is the place to be. There is a dark side to science.
- Postdocs have low pay and high stress. Industry pays better, especially as you move into middle management.
- Working with grants is difficult, there is pressure to produce and get grants.
- Takes too long (6 to 8 years) to get degree.

UCSF participants were asked what factors affected their career choices.

- Advocates for community. Success leads to money. This will enable him to give something back to the community. But still wants financial security to enable to fund community projects.
- Was surprised how hard people are willing to work. This is part of the UCSF culture. Life beyond work is important.
- Work level is shocking and it surprised him. Everybody is doing science all the time.
- Wants a life and wants to get excited about things outside of science.
- Secondary school teaching was very hard. Had to be on call all the time. Although it has less flexibility, he likes it.
- Research first – teach later; you can't do both. You can't achieve balance across time, but you can over time.
- UCSF is very stimulating, small college will be broader but less intense. Secondary school compensation is too low. Teachers should be given better tools.
- You must raise the pay levels to make secondary school teaching more attractive.

- Summer money is an issue with the public.
- Looked for peer interaction in secondary school teaching, not as much as hoped for or desired.
- Likes the level of interaction with the grad students.

Right now, what do you think are your chances of realizing your aspirations? What are the obstacles to realizing these aspirations?

***Rutgers/UMDNJ*:**

- The job market is very difficult now for tenure-track faculty positions.
- There are hundreds of applications for positions, even at small institutions.
- Only 5 percent of Ph.D.s will obtain tenure-track faculty jobs [check for actual percentage]
- A research career is dependent on your research outcomes.
- All Ph.D.s have a strong foundation in scientific concepts, so luck is a factor in generating a hot research project at the right time.
- Getting one of these positions is also dependent on who you know, having the right connections.

Duke:

- The psychology job market is horrible but the neurobiology market is doing really well so he has retrained himself and hopes his chances for a research position at a major research university are good.
- In zoology, Duke grads have a better than average chance of attaining research jobs; a postdoc is needed first.
- Mathematics: The job market is awful and the dual career issue has made matters worse. Her career is on hold while her husband finishes his Ph.D. She once considered the academic faculty track but doesn't believe it's likely.
- Environmental education: already doing what she wants to do.
- Undergraduate teaching: reasonably confident about obtaining a position at a liberal arts college.

***UT Austin*:**

- Math job market is depressed, must consider alternatives, but you are not taken seriously if you look outside of academe I am developing a product for my PI who has his own start-up company. He will support me in getting a job.
- The old boy network hurts females; fecundity is a liability, the potential for spousal mobility is a problem; the government has come to terms with this, while industry and academe have not.
- People hit the ceiling in industry.
- Not everybody can teach and do research.
- Working under grants is tenuous, when the grant runs out, research is sidetracked.
- Family is important and the lab requires too much time.

***UCSF*:**

- All 10 participants are involved in the SEP program that gets them involved in high schools, but not as teachers. In addition, all students are required to serve as a TA at some time in their student careers. Since two want to teach (one at a community college and one at a secondary school) and three others are considering teaching at the community college level, many of their comments focused on teaching.
- What is the next step after the postdoc? Wants to make a difference and may teach to pay back.
- Not much teaching experience in grad school. Teaching is discouraged by the principal investigators (PIs), so you must gain teaching experience on your own.
- Because there are no undergraduate students, faculty who are interested in teaching are not attracted to UCSF. This makes it difficult for students who want to teach.
- The PIs are not happy about the TA teaching requirement but support SEP activities.
- Took time off from graduate school to teach at secondary school to gain experience (summer only, taught a summer school session). This had a negative impact on his career, but it was worth it.
- It is not possible to take time off, fields are too competitive, too much is happening too fast
- It's hard to even take a vacation, the PIs want you in the lab all the time.
- Grad students (and other faculty) in other areas have time off – this is frustrating; there is a penalty to take time off.

Think back to when you first started college and when you first started graduate school. What were your career aspirations at each of these times? If you have changed your aspirations, what caused you to change?

***Rutgers/UMDNJ*:**

- One participant had originally been pre-med and decided that, instead of working with people; she preferred working in a lab doing research [not a good candidate for teaching?]
- One participant had previously earned a degree in special education and was turned off by both the teacher education courses (superficial use psychological theories misconstrued in the first place) and their application in the educational environment.
- As noted above, about half suggested they have always aspired to positions in academic research, an aspiration reinforced by their mentors and the graduate school culture.
- One person said she once aspired to academia but was now considering industry.
- Two said that they had considered secondary school teaching as an option, but one had fully rejected the notion and the other (whose husband has become a secondary school teacher) said that she had decided instead on starting her own contract research business.
- One participant said that since his wife earns a good income he has the flexibility of considering options (such as teaching) he might not otherwise have considered.

Duke:

- Postdoc: had subtly shifted from psychology to neurobiology because of the job market; he said he's working on the same issues, just with different tools.
- Math: had aspired to being a great mathematician, but now wouldn't want to be a math researcher; the job market is difficult and after so many years of doing research by herself she wants to do something less lonely that involves people.

- Zoology: don't like the priorities you have to set to make it as an academic researcher so has focused now on undergraduate teaching.
- Environmental education: changed from environmental management to environmental education; then had considered secondary school teaching and had a chance to work in a high school biology classroom, but now is more interested in education programming and teacher education in environmental studies. (NC recently passed a law requiring that earth and environmental sciences be taught in secondary schools. Most science teachers are not trained in this area and would require some training at this point.)

***UT Austin*:**

- Interested in applied biotechnology.
- Wants to go into applied math – thinks there is lots of money in applied fields.
- Wants close ties to industry, wants an industrial focus or a governmental focus.
- Doing research and getting tenure – wants to be in a big academic research center.

FHCRC:

- They got into science through the usual routes, good teachers in high school turned them on to science. Now many wonder what they are going to do "when they grow up." This group was mature and quickly got to the point of discussing secondary school teaching. They recognized that they hadn't been trained to secondary school teachers, but had been trained to be basic scientists. Further, they recognized that most of their professors had never been trained to teach. Although many had worked as teaching assistants as graduate students, they recognized that that experience would not prepare you for secondary school teaching. Many expressed an interest in secondary school teaching and some had ventured into the field.

Have you ever considered a career other than in a university, such as in industry (biotechnology, pharmaceutical) computers, science writing, law school, or teaching?

Rutgers/UMDNJ:

- As noted above, participants have considered: industry, starting independent research business, and secondary school teaching.
- One person mentioned patent law as another career option but did not indicate that he had considered it.
- Participants' perceived positives about jobs in industry were better hours, higher salaries, better benefits, and funds available for undertaking research. "You don't need to secure grants." Perceived negatives were not as independent in conducting research, lose control over research, and research outcomes are held as corporate secrets instead of furthering scientific knowledge.
- Risks: small biotech firms could go under (though if you're an assistant professor and you don't get grants you can go under, too).

Duke:

- As at Rutgers/UMDNJ, there was discussion that (1) graduate programs don't expose students to much beyond academic research, and (2) there is active discouragement of other careers… "if you say you're interested in something else, they cut you off"

- Still, some information does seem to get to these students: in the Math department there was a career symposium that focused on industry and finance, particularly Wall Street. The participant was disappointed that the symposium didn't include options like secondary school teaching and curriculum development that she would be interested in and doesn't know how to get into.

***UT Austin*:**

- Liked teaching, but major professor wanted her in the lab all the time.
- Likes small biotech companies, this enables one to acquire skills and move from lab to lab.

***UCSF*:**

- Seminar series on practice of science career center, SEP. Bruce Alberts is a good role model for alternative careers.
- Got scooped – will get scooped – PIs are afraid to tell students the reality of science. When you start grad school you never think about the end.
- When you enter graduate school, you focus on your thesis project, not about careers.

FHCRC:

- Industry – has some bench science.
- Writing, but I just don't know how to get into it.
- Traditional careers are often at odds with family formation.
- Small college teaching presents similar problems.
- Supporting the lab is a major issue.

Have you ever considered a career as a secondary school science or mathematics teacher? If yes, why? If not, why not?

***Rutgers/UMDNJ*:**

- Five out of 10 participants had considered secondary school teaching on some level: 2 of these individuals have parents who are secondary school teachers which was responsible, in part, for the thought crossing their minds.
- One person said he had never considered the option until he heard about the focus group; he was focused on obtaining a research position.

Duke:

- Four of the five participants had considered or been involved in K-12 teaching:
- One had previously taught autistic children, but was now seeking a position in zoology research.
- One had taught secondary school English in Japan, but was now focused on obtaining an undergraduate teaching spot.
- One had experience in the last year with student teaching in secondary school biology, but was now working in teacher education in environmental studies.
- One Ph.D. in math would like to pursue secondary school teaching as an option and thinks it wouldn't be too different from teaching math to college freshmen.
- One definitely was not interested.

***UT Austin*:**

- Four had considered teaching. One ruled it out because he wanted to be less people oriented and more research oriented. Another likes math too much to teach 10th grade algebra forever

***UCSF*:**

- One participant is now considering secondary school teaching as an option. Others have considered, but would prefer teaching undergraduates.

FHCRC:

- This group quickly got to the point of discussing secondary school teaching. They recognized that they hadn't been trained to be secondary school teachers, but had been trained to be basic scientists. Further, they recognized that most of their professors had never been trained to teach. Although many had worked as teaching assistants as graduate students, they recognized that that experience would not prepare you for secondary school teaching. Many expressed an interest in secondary school teaching. Some had ventured into the field.

What might be attractive about such a position?

***UT Austin*:**

- Good secondary school teaching is better than bad college teaching.
- Teaching at a super high school could be a turn-on. Respondent wants to teach advanced courses to maintain interest in math alive.
- There are positive social issues associated with secondary school teaching. Respondent wants to give something back.

FHCRC:

- The summer option to do research is attractive.
- The salary advantage of the PhD is good.
- You have the ability to increase the knowledge level in secondary school.
- Time-limited teaching experience would be interesting. The 5-year option is too long, a 1- or 2-year option would be appropriate. The first year would be hell, the second year would be productive.
- The best time to capture Ph.D.s for secondary school teaching is right after graduation.
- If you put an ad in *Science* or *Nature* with an established program to recruit Ph.D.s scientists to secondary school teaching, it would work. Lots of students would sign up.

What makes this position unattractive to you?

***Rutgers/UMDNJ*:**

- Getting a Ph.D. involves specialized, advanced research while the science courses one would teach in secondary school are so basic: I wouldn't be teaching what I learned over 8 years as grad student and postdoc, and I would feel as though I wasted my Ph.D. In response to this, another participant said that it would depend on what the course is. If it's basic bio, then it would be boring; but it would be interesting if you could teach students how to set up and carry out experiments. Another said that they don't do that even with undergraduates. A final participant suggested that they ought to teach this to students at both the college and high school levels.

- Worried that there would be conflict between teachers with Ph.D.s and teachers who don't have Ph.D.s
- Not sure I would have a good sense of how to measure the progress of secondary school students.
- Wouldn't be able to deal with high school students: if you work in a high school you don't just teach, you have to deal with teenagers.
- Teaching (even at the university level) gets in the way of the research you want to do. Research is fun.
- Teaching in a secondary school would pull you out of access to the scientific community (other scientists, seminars, etc.).
- The problem is not the kids, not the salary, not the loss of research, it's the other teachers I'd have to work with.

Duke:

- Couldn't teach psychology in secondary school, but could teach in college/university where it is one of the largest majors.
- Couldn't teach specialized, fun courses. You'd have to teach basic H.S. biology
- Lack of ability to control the curriculum.
- Don't have the flexibility to be creative…have to follow the textbook
- Lose touch with other scientists.
- Hard to teach when you have 34 students.
- Problems of public schools: discipline, violence, drugs. Are students interested in learning? Other teachers have lost the passion?
- Worried that I wouldn't be up to the task given these problems.
- Kids hate math.

***UT Austin*:**

- Student quality is low, social promotion is prevalent.
- Students are "grubbing for grades."
- Likes math too much to just teach 10th grade algebra.
- Students are not motivated to do well. The motivation of students is more important than their actual knowledge.
- Professors said that with a Ph.D "he could do better" than secondary school teaching

FHCRC:

- No respect given to secondary school teachers.
- Stigma attached to secondary school teaching (by academe).
- Structured curriculum, no creativity.
- Lots of time is invested for secondary school teaching. The PhD is over educated to teach secondary school.

- If you leave academe for secondary school teaching you can't reenter into academic science. Secondary schools don't respect the value of a Ph.D.
- Nobody goes to a major university because they want to teach, but because they want to do research.
- Change is an uphill battle.

What are the obstacles to pursuing a position as a secondary school teacher?

***Rutgers/UMDNJ*:**

- Taking teacher education courses that are "dumbed down" and not rigorous.
- The administrations in school districts don't want to pay for Ph.D.s when they can hire M.A.s and B.A.s.
- Note: participants said their own peers would be supportive, but that their professors would not.

Duke:

- Don't know how to obtain a position in a secondary school.
- School systems want to hire B.A.s and not even M.A.s because they don't have to pay as much.

***UT Austin*:**

- Dealing with the local school bureaucracy is a major problem. Kids are manipulative and teachers are under a constant pressure to defend themselves against the kids. There are feelings that the bureaucracy backs the kids and hinders the teachers.
- Once you get out of academia you can never come back. Most teaching jobs at colleges are really research jobs with some teaching responsibility.

FHCRC:

- PIs want you to stay in the lab.
- There is resistance by the PIs for students to have TAs. This makes it tough for graduate students to get any teaching experience.

What about teaching gifted students or in magnet schools?

***Rutgers/UMDNJ*:**

- That would be better.
- Might consider if it was part-time and I could do research part-time.
- Magnets: positive that the parents and kids would both want Ph.D.s teaching in these schools.

Duke:

- Would be better since there might be more resources.
- Students at a magnet might be more motivated and thus have the energy to undertake independent research projects.
- School of Math and Science: these kids don't need their help; the socially responsible thing to do is to teach lower-level students.

- A person who goes for a Ph.D. is usually an idealistic, creative person ... teaching at a magnet school would be a cop-out.
- Magnets: students aren't necessarily at a higher level; magnets are often created to address integration.

FHCRC participants were asked what incentives would get them to consider secondary school teaching.

- Get teaching certificate as a result of supervised teaching.
- Have teaching count as part of the NIH payback.
- Get sent to professional meetings.
- Free subscription to *Cell.*
- Include funding for lab expenses as part of salary package (this will let them conduct interesting experiments for the classes).
- If you really want to attract Ph.D.s to secondary school teaching, let them take two years to try student teaching with a stipulation that if they want it, they will have a guaranteed two-year postdoc, funded by NIH. This means that if they take the risk of leaving research for secondary school teaching and it doesn't work out, they will have an opportunity to get back into research.

Would you consider taking such a position for a five-year commitment?

***Rutgers/UMDNJ*:**

- Can't take a period of time off and then get back on the research track.
- Summer research might not help: Howard Hughes Medical Institute has a program to bring teachers into the labs of Hughes investigators and it takes them the whole summer just to get geared back up again.

Duke:

- If you take two years off to teach, you won't get back on to the research track.

***UT Austin*:**

- If there were no penalties, a two-year postdoc in secondary school teaching would be a good idea.

***UCSF*:**

- Yes.
- Yes, and include training as part of a post doctoral experience.
- Yes, and get credentialed as part of process
- There needs to be a change in faculty attitudes.
- The term "alternate careers" needs to be changed to careers.
- More people go into the alternatives than into academics.

- Give a masters degree as part of the graduate process – say at the time you are admitted to candidacy – so that people can step out early and still get recognition for what they have done.
- This is being considered
- Develop a web site with information about secondary school teaching.
- The Ph.D. commands respect, we should be able to use it to modify curriculum and decrease classroom size

FHCRC:

- Five years is too long, but would consider two years to try student teaching with a stipulation that if they want it, they will have a guaranteed two-year postdoc funded by NIH. This means that if they take the risk of leaving research secondary school teaching and it doesn't work out, they will have an opportunity to get back into research.

For teacher certification, most states require teacher education courses and a passing score on basic skills and/or subject matter examinations. Would you consider secondary school teaching if you could satisfy the teacher education requirement by (see handout)?

***Rutgers/UMDNJ*:**

- One person said that he wouldn't want to take teacher education courses while in grad school on top of everything else and that if he did decide early on in grad school to become a teacher he wouldn't bother to get a Ph.D. Rather, he'd leave with an M.A.
- Another said that he wouldn't want to take courses while starting a teaching job. That wouldn't be fair to the students to have an untrained teacher.

Duke:

- One responded that he wouldn't take teacher education courses in grad school; it would tag you as a dropout.
- Another said he already had teaching experience from five years as a TA. So he didn't see the need for teacher education courses.
- Another said that in psychology we don't understand how people learn very well. He doubted that education departments really know.
- The MS/MAT participant said that the adolescent psychology taught in teacher education courses was not very useful; these courses were useful only to the extent that they got you "in the mode" of thinking about how to teach. What was most valuable was time spent in the classroom.
- Participants would just want to pass an examination. One participant suggested that certification should be granted based on teaching time and a passing score on an exam
- One person said that instruction in "cooperative discipline" would be helpful, but that this could be taught in a one-week workshop and wouldn't require a full-scale course. Another agreed that they'd need direction on how to control students.

***UT Austin*:**

- State rules are a major obstacle to getting Ph.D.s into secondary school teaching.

- The group had a very low opinion of college of education departments and courses, especially at large state universities. They believed that education students were the worst of any in higher education.
- The only acceptable path to certification is option 4 (taking exam without taking extra courses). Option 1 is acceptable only if you make the decision to go into teaching early.

***UCSF*:**

- Only option 1 was acceptable to the students.
- Exam does not make you a teacher. You must learn how to teach. Good teachers do not do it because they can't do anything else.
- Taking teaching classes while in grad school is part of core curricula.
- The SEP experience was good. I did not get support from other teachers, but I didn't need it.
- SEP experience was great, but I would like to have some child psychology classes and mentoring support. Wants an established, structured program of mentoring.

***FHCRC*:** (Many of the focus group members had had some level of involvement in secondary school teaching)

- It is important to get some course work and student teaching before taking the certification exam.
- Teaching courses are good for "problems," but teaching is best learned by doing.
- An apprenticeship program is needed, (e.g., team teaching for two or three persons in a secondary school accompanied with meetings with a master teacher periodically to work out problems). These mentors and master teachers should be compensated for their work.
- I would feel much more comfortable with a mentor.
- The group was worried about "hordes of untrained people ascending upon high schools to teach". The majority of the group had a problem with teaching without training.

What career options would you be willing to consider based on the following salaries for professional positions and your perception of the chances of your obtaining one of these positions? (see state-specific handouts)

***Rutgers/UMDNJ*:**

- The salary data presented did not appear to sway the participants in any way.
- One person remarked, looking at the data, that it was sad what scientists are paid regardless of the position.

Duke:

- When confronted by the salary data, the participants were unanimous that factors other than salary determined what they wanted to do: (1) flexibility to schedule the day as they wanted, (2) or have a workday that stopped at 6 p.m., (3) be able to publish, and (4) have a job that allowed arrangements such as splitting time teaching secondary school and teaching undergraduates at Duke (as someone now does).

***UT Austin*:**

- The teacher's salaries in Texas are very low. All respondents commented on the low pay. The consensus was that industry pays best.

- Teachers are rated by pupils test scores, but the problems start long before the teachers come in contact them.
- The low salaries reflect Texas' attitudes towards teaching.
- Other aspects of compensation, besides salary, that might sway participant's decisions were: loan repayments; being able to work independently of current curriculum development, (e.g., design your own course; make changes in the program), Ph.D. salary differential; and freedom from teacher's unions.

UCSF:

- Thinks starting salaries for secondary school teacher is OK.
- Postdocs are underpaid. Grad students are cheap labor
- Salary is a big consideration.
- Best salaries are in industry.
- Note: San Francisco is very expensive, the students know this and understand impact of the cost of living on career choice. Six came to the UCSF because they wanted to come the Bay area, while four wanted to come to UCSF specifically.
- Other factors that would influence career: select a career with time off for family; geography is important, but it limits choices.
- All received full stipends so loan repayment was not an issue.

FHCRC:

- The salary level for secondary school teaching is better than for a postdoc.
- Salary becomes more important as the one grows older.
- Family considerations become very important, and dictate the need for more and stable money.

What characteristics of your work environment would be especially important to you in a secondary school teaching position?

Rutgers/UMDNJ:

- Having an ability to do research was important for several participants.
- Would need to have a connection to the scientific community and have Internet access.
- Would need to address the issue of tension between Ph.D. teachers and other teachers.

Duke participants were asked to comment on the idea of Ph.D.s having the chance to conduct research while they taught secondary school:

- Time is an issue. You can't be cutting edge, if you don't have the time or the resources, or the grad students to do the lab work.
- Hard to stay in touch with the scientific community if you're not in research full-time
- You can't accomplish anything in a summer.
- Would have a hard time making deadlines for filing conference papers.

***UT Austin*:**

- More separation between teachers and syllabus.
- Curriculum development.

***UCSF*:**

- It would help if there were more respect for teachers.
- Increase the variety of courses.
- Time off.
- Salaries are lower in private schools, but the level of respect is higher.
- Classroom size is important. A class with 30 kids is scary.
- Private school may be a way of hiding from reality.

Would you consider taking a position with a school district or state department of education in science or mathematics curriculum development? If yes, why? If not, why not?

***Rutgers/UMDNJ*:**

- Note: the participants seemed never to have considered this kind of position before. They seemed to find some appeal to it.
- Participants had a very positive reaction to the idea—offered by one of their peers—of taking positions as teachers of science teachers (as opposed to becoming science teachers themselves).

Duke:

- One respondent was very enthusiastic about the idea: her dream job would be to teach 9th grade math one year and develop an appropriate 9th grade math curriculum as she went; then move on to 10th grade math and so on until she had developed a full curriculum for grades 9-12.
- Another, who would not consider secondary school teaching, said he might consider curriculum development as an alternative to undergraduate teaching.
- The MS/MAT participant said that she is already working in this area.
- When asked about teacher education (i.e., "training the teachers") there was interest, but not as much as for curriculum development.
- Again, the MS/MAT participant who is also working on curriculum development is already doing this. She said, "The most exciting part of teacher training is taking the teachers out and doing a lab that they can take back to their classrooms and use with their students."
- The participants at Duke suggested other ways Ph.D.s could help improve science literacy in the United States were to have graduate students satisfy their grad school teaching requirements by teaching in secondary school rather than serving as a TA for undergraduate courses, and work to in museum programs with K-12 students.

***UT Austin*:**

- Had already done that, enjoyed it very much.

- Good idea, especially developing experiments for secondary school teachers.
- Yes, because of the poor preparation of secondary school students, need to increase standards.

***UCSF*:**

On curriculum development:

- Yes, interesting
- Yes, "but it's hard to write magic into a curriculum." Teaching depends on individual skills.
- Yes, seriously thought about it – have some similar work through SEP. Curriculum development needs to be integrated with classroom experience or nobody will listen to you
- Yes, but where do you start.
- I know so little about secondary school that I would be intimidated.
- No standards. Could be limited.
- Curriculum development is fine, but the magic is the interactions with the teachers.

On training teachers:

- Yes, designed experiments in cooperation with secondary school teachers
- Yes, summer program with partners to teach K-5 teachers. Teach them how to use kits that explain science.
- Partnership was great – she learned teaching!
- Long-term (informal) partnership was valuable.

FHCRC:

- I didn't know that this was a career option
- Curriculum development is good because if effects the most people.
- They can best use their experience in curriculum development.
- Ph.D.s are very focused on topics, but they can see beyond their immediate areas of interest.
- You need some teaching experience to work in curriculum development.
- Need a three-way team of scientist, educator, and curriculum development administrator to be effective.
- The group had some problems with the level of secondary school textbooks. They see curriculum development lagging way behind what is happening is science. They believe that if scientists were part of the curriculum development team, the level of science in secondary school classes would be brought up to date.
- I'd like to work as a teacher trainer, the Hutch as one (University of Minnesota has one), that would be very interesting.
- The SEP program works too.
- However, labs are unwilling to accept active involvement in a SEP program.

What might be attractive about such a position?

***Rutgers/UMDNJ*:**

- Would have a broader impact than just teaching.
- Could influence the students without having to be a teacher.

- Would especially like it if it involved teaching the teachers, because science teachers are not in touch with researchers (this was a popular sentiment and option!)
- Would enjoy looking at how the different pieces of the science curriculum (elementary, middle, and secondary school) fit together.

Duke:

- Have "already been doing it" in the sense that she has designed courses for freshman math.
- Would enjoy thinking about the subject (mathematics) globally, putting the pieces together.
- It would be really rewarding to see kids achieve because you've figured out how a subject should be taught.
- Once you've gotten to the M.A./Ph.D. level, you have skills and interest to make a difference looking at the "big picture."

***UT Austin*:**

- Already doing it.
- It's better if it's done collectively, there's an incentive for the teachers.

What might be unattractive about such a position?

Rutgers/UMDNJ: No negatives mentioned.

Duke

If you're working in the North Carolina Department of Public Instruction, you're far removed from students and teachers.

- Local school boards often have substantial power: your well-thought-out curriculum might be torn to shreds in implementation.
- Low tolerance for bureaucracy.
- Ph.D. training doesn't prepare you for administrative jobs.

***UT Austin*: Dealing with the school bureaucracy is a major problem.**

- It would be difficult dealing with teachers who are a lot older than I am.

Appendix C

Survey of Graduate Students and Recent Ph.D.s: Instruments and Protocols

NATIONAL RESEARCH COUNCIL
Survey of Graduate Students and Recent Ph.D.s

1. What is your current field of science or mathematics? *(please fill in)* _____________

2. What was your undergraduate major? *(please fill in)*___________________

3. Are you currently a
   - ❑ Graduate student
   - ❑ Postdoctoral fellow
   - ❑ Other (*specify*) __________________

4. How long have you been in this position? __ __*years* __ __ *months*

5. In the table below, we are interested in learning about your next step in employment or training and your career aspirations. *Please indicate these in columns B, C, and D as follows:*

   In Column B: Please mark (X) the ONE category that best represents your immediate next step in employment or further training following your current position.
   In Column C: Please mark (X) the ONE category that best represents the position to which you aspire in your career at this point in time.
   In Column D: Please mark (X) ALL of the positions you have considered in thinking about your long-term career plans.

A Position	B Immediate next step in employment or training (*mark ONE*)	C My career aspiration (*mark ONE*)	D Positions I have considered (*mark ALL that apply*)
Postdoctoral fellowship			
Faculty in a research university/ institute or medical school			
Faculty in a liberal arts college			
Faculty in a community college			
Other position in a college, university, research institute, or medical school			
Position in K-12 science/mathematics education, teaching, or leadership			
Position in a non-profit organization			
Position in government			
Position in industry			
Self-employed			
Other (specify)			

6. What is the annual salary you receive for your current position (include fellowships, stipends for teaching, and research assistantships)? (*please estimate salary*) $

7. What annual salary do you expect for your immediate next step in employment or training (the position you selected in column B)? (*please estimate salary*) $

8. What annual salary would you expect in the position to which you aspire if you held it seven years from today (position selected in column C)? (*please estimate salary*) $

9. Please indicate all of your teaching experiences. *(mark all that apply)*

- ❑ Elementary school faculty (grades K-6)
- ❑ Secondary school faculty (grades 7-12)
- ❑ Volunteer work in K-12 schools
- ❑ Junior or community college faculty
- ❑ Teaching assistant in a college or university
- ❑ College or university faculty
- ❑ None
- ❑ Other (specify) ___________________

10. What was the year of your most recent teaching experience? *year:* 19 __ __

11. Have you ever actively considered taking a position teaching science or mathematics in a secondary school (grades 7-12)? ❑ Yes ❑ No → *Whether you marked yes or no, please proceed to question 12*

12. *Please mark yes or no to each of the phrases below completing this sentence:* "I would consider taking a position as a secondary school science or mathematics teacher if…"

Yes	No	**Summer Research**
❑	❑	I would be guaranteed a summer fellowship at a research lab including travel expenses
❑	❑	I would be guaranteed a summer fellowship in a research lab
❑	❑	I would be able to compete for a summer fellowship in a research lab
❑	❑	I would be able to participate without compensation in a research lab during the summer
❑	❑	I would be free to do as I pleased during the summer months

Yes	No	**Participation at Scientific Meetings**
❑	❑	I would be given funding and time to attend two or more scientific meetings during the school year
❑	❑	I would be given funding and time to attend one scientific meeting during the school year
❑	❑	I would be given time to attend scientific meetings during the school year
❑	❑	I would be given time to attend one scientific meeting during the school year
❑	❑	I would have no opportunity to attend scientific meetings during the school year

Yes	No	**Assistance in Curriculum Development and Teaching**
❑	❑	A specially selected, experienced teacher would mentor me during my initial 2 years of teaching
❑	❑	A specially selected, experienced teacher would mentor me during my initial year of teaching
❑	❑	A specially selected, experienced teacher would mentor me during my first semester of teaching
❑	❑	A specially selected, experienced teacher would be assigned to mentor me on an as-needed basis
❑	❑	I would have no formal assistance from an experienced teacher

Yes	No	**Certification Requirements**
❑	❑	Teaching proficiencies could be demonstrated on the job
❑	❑	I could demonstrate proficiencies prior to becoming a teacher, through prescribed formal courses or experiences based on my needs
❑	❑	I could receive teacher certification by taking an intensive summer course in education
❑	❑	I could receive teacher certification by taking one year of education courses
❑	❑	I could receive teacher certification by taking more than one year of education courses

Yes	No	**Duration of Teaching Commitment**
❑	❑	I would accept a 1-year teaching commitment
❑	❑	I would accept a 2-year teaching commitment
❑	❑	I would accept a 3-year teaching commitment
❑	❑	I would accept a 4-year teaching commitment
❑	❑	I would accept a 5-year teaching commitment

13. *Please mark yes or no to each of the phrases below completing this sentence:* "I would consider taking a position as a secondary school science or mathematics teacher..."

	Yes	No
A. if I were awarded a prestigious national fellowship that would provide training, placement, and special opportunities for networking with peers, as well as cover my living expenses during the training period.	❑	❑
B. if I were provided with the option of a two-year guaranteed postdoctoral research fellowship at the end of a two-year teaching commitment, in addition to the national fellowship described in 13A above.	❑	❑
C. if I were provided with the option of a two-year guaranteed postdoctoral research fellowship at the end of a two-year teaching commitment, without the national fellowship described in 13A above.	❑	❑
D. if financial assistance were available only for the educational expenses incurred in the process of gaining certification.	❑	❑
E. if I were supported during my teaching by a regional or university-based science teaching resource center that provided science kits, loaned laboratory equipment, and organized field opportunities for science experiments in which students could participate.	❑	❑
F. if I were able to teach in a specialized science and technology high school.	❑	❑
G. if one year of my student loans were forgiven for each year of employment in a full-time teaching position.	❑	❑
H. under no circumstances.	❑	❑

14. *[If you are a graduate student]* Would you consider taking a position as a secondary school teacher following receipt of a Masters degree from your current program?
❑Yes ❑No

15. What would you require as a beginning annual salary if you were a secondary school teacher?
(*please estimate salary*) $

16. What would you require as a salary after seven years as a secondary school teacher?
(*please estimate salary*) $

17. Would you consider other full-time positions in K-12 science and mathematics education?
Please answer yes or no for each position

Yes	No	
❑	❑	Curriculum development (devising or refining units of study by grade level)
❑	❑	Professional development (teaching science or mathematics teachers)
❑	❑	Science or mathematics specialist for a school district
❑	❑	Science resource teacher for an elementary or secondary school
❑	❑	K-6 science or mathematics teacher
❑	❑	Science education partnerships (university- or industry-based expertise for K-12 classrooms)
❑	❑	Science specialist in a regional or university-based science education resource center
❑	❑	Working in a science museum, environmental science center, or similar institution

18. Do you have any close family members who are or have been elementary or secondary school teachers? ❑Yes ❑No

19. What would be attractive to you about secondary school science or mathematics teaching or other positions in secondary school science or mathematics education? ____________________________
____
____
____

20. What would be unattractive to you about these positions?______________________________
____
____
____

21. Are you
- ❑ Male
- ❑ Female

22. What is your year of birth?
year: 19 __ __

23. What is your marital status? *(mark one)*
- ❑ Single, never married
- ❑ Married
- ❑ Living in a marriage-like relationship
- ❑ Separated, divorced, widowed

24. Not including yourself, how many dependents do you have—that is, how many others receive at least one half of their support from you—under age six or age six and over?
Dependents under age six __ __ *(number)* *Dependents age six or over* __ __ *(number)*

25. Do you anticipate that the number of your dependents will increase in the next two years?
❑ Yes ❑ No

26. What is your citizenship status?*(mark one)*
- ❑ United States citizen, native
- ❑ United States citizen, naturalized

→ *go to question 28*

- ❑ Permanent resident of the United States
- ❑ Temporary resident of the United States

→ 27. Of which country are you a citizen?
(country:) ____________________

28. Are you Hispanic?
- ❑ Yes
- ❑ No

29. What is your racial background?
(mark all that apply)
- ❑ American Indian or Alaskan Native
- ❑ Asian or Pacific Islander
- ❑ Black or African American
- ❑ White

30. How much money do you currently owe that is directly related to your undergraduate and/or graduate education (tuition and fees, living expenses and supplies, transportation to and from school)?
(*please estimate debt level*) $ ____________

Please provide any additional comments you may have. Use additional paper as needed.

FOCUS GROUP PROTOCOL

NATIONAL RESEARCH COUNCIL

Office of Scientific and Engineering Personnel Center for Science, Mathematics, and Engineering Education

Attracting Science and Mathematics Ph.D.s to Secondary School Teaching

Focus Group Questions

Tell me about your current position? What is it and how long have you been in it? [This is an icebreaker: participants will be grad students or postdocs]

What are your career aspirations?

What do you find attractive about the career you aspire to? Unattractive?

Right now, what do you think are your chances of realizing your aspirations?

What are the obstacles to realizing these aspirations?

Think back to when you first started college and when you first started graduate school. What were your career aspirations at each of these times? If you have changed your aspirations, what caused you to change?

Have you ever considered a career other than in a university, such as in industry (biotechnology, pharmaceutical) computers, science writing, law school, or teaching?

What is attractive about these careers?

Probes [Less time in postdoc]

[Better pay and benefits]

[Increase in authority]

[No need to write grants]

[Job stability]

[Can have a life]

What is unattractive about these careers?

Probes [Out of the academic research loop]

[Need more time as a postdoc]

[Not interested in applied research or teaching]

[Am not interested in administration]

[Wouldn't want someone telling me what to do]

What are the obstacles to pursuing these alternatives?

Probes [Lack of jobs]

[Loss of identity]

[Easy to get taken-in]

[Risks associated with working for start-up companies]

[Would need substantial training (for computers, law, teacher education]

Have you ever considered a career as a secondary school science or mathematics teacher? If yes, why? If not, why not?

What might be attractive about such a position?

Probes [Enjoy teaching]

[Enjoy working with young people]

[Would enjoy teaching in a magnet school for science & technology]

[Would be productive way to use my science/math knowledge]

[Would enjoy contributing to the improvement of science and math education in the United States]

[Rediscover general interest in science]

[Community research projects]

[Development of pedagogical skills]

[Potential summer opportunities: teaching, research, vacation]

[Nice alternative to research track and would get me out of postdoc limbo]

What makes this position unattractive to you?

Probes [Am only interested in basic research]

[Material is boring]

[Loss of status]

[Peer disapproval]

[Colleagues are not as stimulating]

[Don't want to work with young people]

[Don't like teaching]

[Would have to manage a classroom and discipline students]

[Would need additional teacher training]

[Low salary compared to university-level academics]

What are the obstacles to pursuing a position as a secondary school teacher?

Probes [Don't know where the jobs are]

[Need state certification]

[Would need extra training to learn to teach]

[Would only work in magnet schools or schools for gifted children]

[May get kicked-out of graduate program if I leave the research track]

[Can't afford it because of student debt—and other alternatives careers would pay more]

[Facing potential rejections by colleagues in the secondary school (either faculty or administrators]

Would you consider taking such a position for a five-year commitment?

Probes [5-years is too long, I'd never be able to get back into basic research]

[5-years is enough time]

[Would only do this as a permanent career change]

[Would only consider it if I were associated with a research program]

What incentives or special arrangements would be required to attract you to a secondary school teaching position? *Follow-up probes:*

1. Certification:

What is your perception of the certification process for secondary school teaching in your state?

For teacher certification, most states require teacher education courses and a passing score on basic skills and/or subject matter examinations. Would you consider secondary school teaching if you could satisfy the teacher education requirement by: (see handout)

2. Compensation:

What career options would you be willing to consider based on the following salaries for professional positions and your perception of the chances of your obtaining one of these positions? (see state-specific handouts)

What other aspects of compensation, besides salary, might sway your decision?

Probes [Summer fellowships for research]

[Student loan repayment]

[Reimbursement for relocation costs]

[Benefits]

3. Work environment: What characteristics of your work environment would be especially important to you in a secondary school teaching position?

Probes [Type of school (e.g., science/math magnet)]

[Age of students]

[Ability to job share]

[Ability to have job security]

[Availability of lab equipment or science kits]

[Special summer programs (for teaching, professional development, research)]

[Ability to contribute to curriculum development]

[Ability to do research]

[Other]

Would you consider taking a position with a school district or state department of education in science or mathematics curriculum development? If yes, why? If not, why not?

What might be attractive about such a position?
Probes [Administrative responsibility]

[Actions and decisions have big impact]

[Can set policy]

What might be unattractive about such a position?
Probes [Completely removed from research]

[Would waste time spent in graduate school]

[No interest in administration or policy work]

[Would rather work with scientists]

What are the obstacles to pursuing such a position?
Probes [New and difficult network to enter]

[Requires new skills and abilities]

[Others would not accept me because I'm primarily a scientist, not an educator

[Would be frustrating to rise through the ranks]

ATTACHMENTS

Teacher Education Requirements

For teacher certification, most states require teacher education courses and a passing score on basic skills and/or subject matter examinations. Would you consider secondary school teaching if you could satisfy the teacher education requirement by:

[For graduate students]

- Taking teacher education courses during graduate school to obtain certification while completing Ph.D. requirements?

[For all]

- Taking teacher education courses after your Ph.D. award and before you started teaching?
- Completing teacher education courses within a specified time period while you begin secondary school teaching?
- Passing an examination without having to take extra teacher education courses?

New Jersey

Position	Salary
Average beginning salary for teacher in New Jersey	$28,039
Beginning salary for teacher with Ph.D. in Paterson, NJ	37,485
Maximum salary for teacher with Ph.D. in Paterson, NJ	69,076
Beginning salary for teacher with Ph.D. in Jersey City, NJ	41,600
Maximum salary for teacher with Ph.D. in Jersey City, NJ	75,500
Stipend for postdoctoral fellow with 0 years experience	26,256
Stipend for postdoctoral fellow with 7 years experience	41,268
Average salary for instructor at Bergen Community College	38,900
Average salary for assistant professor at Bergen Community College	48,800
Average salary for full professor at Bergen Community College	78,400
Average salary for instructor at Rutgers University (New Brunswick)	32,800
Average salary for asst. professor at Rutgers University (New Brunswick)	50,600
Average salary for full professor at Rutgers University (New Brunswick)	94,800

North Carolina

Position	Salary
Average beginning salary for teacher in North Carolina	$21,136
Beginning salary for teacher with Ph.D. in Durham, NC	28,666
Maximum salary for teacher with Ph.D. in Durham, NC	49,588
Beginning salary for teacher with Ph.D. in Charlotte, NC	29,046
Maximum salary for teacher with Ph.D. in Charlotte, NC	51,306
Stipend for postdoctoral fellow with 0 years experience	26,256
Stipend for postdoctoral fellow with 7 years experience	41,268
Average salary for instructor at Appalachian State University	34,300
Average salary for assistant professor at Appalachian State University	41,500
Average salary for full professor at Appalachian State University	60,700
Average salary for instructor at UNC-Chapel Hill	45,600
Average salary for asst. professor at UNC-Chapel Hill	51,200
Average salary for full professor at UNC-Chapel Hill	88,700

Texas

Position	Salary
Average beginning salary for teacher in Texas	$24,079
Beginning salary for teacher with Ph.D. in Austin, TX	26,660
Maximum salary for teacher with Ph.D. in Austin, TX	42,100
Beginning salary for teacher with Ph.D. in Irving, TX	29,725
Maximum salary for teacher with Ph.D. in Irving, TX	49,500
Stipend for postdoctoral fellow with 0 years experience	26,256
Stipend for postdoctoral fellow with 7 years experience	41,268
Average salary for instructor at Stephen F. Austin State University	33,300
Average salary for assistant professor at Stephen F. Austin State University	37,500
Average salary for full professor at Stephen F. Austin State University	55,300
Average salary for instructor at University of Texas	40,200
Average salary for asst. professor at University of Texas	50,600
Average salary for full professor at University of Texas	84,400

California

Position	Salary
Average beginning salary for teacher in California	$26,684
Beginning salary for teacher with Ph.D. in Berkeley, CA	36,083
Maximum salary for teacher with Ph.D. in Berkeley, CA	51,072
Beginning salary for teacher with Ph.D. in San Diego, CA	34,956
Maximum salary for teacher with Ph.D. in San Diego, CA	54,256
Stipend for postdoctoral fellow with 0 years experience	26,256
Stipend for postdoctoral fellow with 7 years experience	41,268
Average salary for instructor at San Diego State University	35,200
Average salary for assistant professor at San Diego State University	45,000
Average salary for full professor at San Diego State University	68,300
Average salary for asst. professor at UC, Santa Cruz	49,200
Average salary for full professor at UC, Santa Cruz	87,200

Washington

Position	Salary
Average beginning salary for teacher in Washington	$23,933
Beginning salary for teacher with Ph.D. in Seattle, WA	29,625
Maximum salary for teacher with Ph.D. in Seattle, WA	49,073
Beginning salary for teacher with Ph.D. in Spokane, WA	32,289
Maximum salary for teacher with Ph.D. in Spokane, WA	49,518
Stipend for postdoctoral fellow with 0 years experience	26,256
Stipend for postdoctoral fellow with 7 years experience	41,268
Average salary for instructor at Eastern Washington University	30,700
Average salary for assistant professor at Eastern Washington University	36,400
Average salary for full professor at Eastern Washington University	55,200
Average salary for instructor at University of Washington	40,800
Average salary for asst. professor at University of Washington	48,100
Average salary for full professor at University of Texas	75,600

PROTOCOL – INTERVIEWS WITH PH.D.S TEACHING IN SECONDARY SCHOOL

NATIONAL RESEARCH COUNCIL

Office of Scientific and Engineering Personnel Center for Science, Mathematics, and Engineering Education

Attracting Science and Mathematics Ph.D.s to Secondary School Teaching

INTERVIEW GUIDE:

PH.D.S TEACHING IN SECONDARY SCHOOL

[**Read to Interviewee:**]

Hello, I'm [Name of Interviewer], a consultant with the National Research Council. I am working on a study that is investigating ways of attracting science and mathematics Ph.D.s to secondary school teaching or curriculum development positions. I hope that you received the letter from the National Research Council last week introducing me to you.

As part of the study, the National Research Council is interviewing a number of science and mathematics doctorates who are currently teaching in secondary schools. We are interested in learning about the process that led you to secondary school teaching and your experiences as a teacher. I expect that this interview will last between 30 and 40 minutes.

[**Questions:**]

1. First, would you tell me something about your educational background?

Probes [Ph.D. institution, field, and year of graduation]

[Experience as a postdoc]

[Employment prior to current position]

2. What attracted you into secondary school teaching?

Probes [Had always wanted to teach

[Enjoy working with children]

[Fruitful way of applying science knowledge]

[Alternative to succession of postdocs]

[Alternative to frustrations of the research track]

[Was unable to get a job at a college or university]

[Took the job for financial and/or family considerations]

3. Were your colleagues or mentors supportive when you made your decision to enter into secondary school teaching? Did you meet with approval or disapproval?

4. Did you encounter any barriers to taking a secondary school teaching position? How did you overcome these barriers?

Probes [None]
[Difficulty finding job openings]
[Getting certified to teach]
[Needed teacher training]
[Adjustment to new working environment]

5. To what extent did financial considerations have an effect on your decision to enter into secondary school teaching?

Probes [Not at all]
[I got higher pay teaching than as a postdoc]
[Wanted benefit package]
[Wanted stable income]
[Concerned about paying off student debts]

6. To what extent did family considerations have an effect on your decision to enter into secondary school teaching?

Probes [None. I was not married or didn't have other family considerations at the time]
[I wanted to settle down and raise a family and wanted more career stability]
[I had a definite location preference and secondary teaching gave me the opportunity to live there]

7. How did you obtain teacher certification? Was this the typical way teachers are certified in your state or was this an alternative certification process? What alternative certification processes are available in your state?
8. Did your State (and/or the State Department of Education) provide you any assistance or pose any obstacles in getting you certified? If yes, how.
9. Would you please describe the school in which you are currently teaching?

Probes [Public or private school]
[Science and technology magnet program or school]
[Urban, suburban, or rural]
[Large or small]
[Skill level of students]
[What kinds of facilities are provided?]

10. What are your relationships with other faculty and with the administration like? Are there other Ph.D.s teaching at the school?

11. Are you still engaged in research? If yes, describe your research activities. If no, why are you not doing any research?

Probes [Summer programs or sabbaticals]

[Linkages with research universities]

[Maintains own lab]

[Networking with colleagues]

[Attends professional meetings]

12. Have you ever had any regrets about entering secondary school teaching? If yes, what were they? If no, do you think that you are a special case?

Probes [Felt loss of status]

[Don't enjoy working in a bureaucracy]

[Lost connection to research community]

[Teaching isn't what I expected]

[Salary and benefits aren't sufficient]

13. What advice would you give to a science/math Ph.D. who is considering secondary school teaching?
14. What do you think about the future of secondary school science teaching? What future do you envision for science and mathematics doctorates entering into secondary school teaching?

PROTOCOL – INTERVIEWS WITH ADMINISTRATORS

NATIONAL RESEARCH COUNCIL

Office of Scientific and Engineering Personnel Center for Science, Mathematics, and Engineering Education

Attracting Science and Mathematics Ph.D.s to Secondary School Teaching

Interview Guide:

High School Principals

Hello, this is [name of interviewer] with the National Research Council. I am working on a NRC project investigating the potential for attracting recent science and mathematics Ph.D.s to secondary school teaching or curriculum development positions. I hope you received the letter we sent you last week introducing you to this project.

As one step in this study, we are conducting a national survey of graduate students and recent Ph.D.s to assess their level of interest in such positions, the obstacles they face in taking them, and the incentives they would need to make the transition to these positions. As another step, we would like to obtain the views of high school principals, district superintendents and state education commissioners on the prospects for placing Ph.D.s in these positions.

1. Do you see recent Ph.D.s as a source of quality science and mathematics teachers in your high school or other high schools in your school district?
2. Why or why not?
3. What do you see as the advantages of Ph.D.s as secondary school science and mathematics teachers?

Probes [Knowledge of subject at the forefront of the field]

[Commitment to subject]

[Ability to convey scientific methods to students]

[They would make fine teachers]

[Would provide an alternative career for them—would anyone say this?]

4. What do you see as the disadvantages of Ph.D.s as secondary school science and mathematics teachers?

Probes [Lack of teaching ability]

[Lack of commitment to secondary education]

[They're trained for and mainly interested in research]

[Unable to thrive in K-12 environment]

[Unable to relate to other faculty]

5. What are the obstacles to attracting and retaining science and mathematics Ph.D.s as secondary school teachers?

Probes [Lack of teacher certification]

[Need for further training in education]

[Lack of interest or commitment on the part of Ph.D.s in teaching]

[Would not be made welcome by other teachers]

[Salary/compensation levels are not sufficient to attract them]

[School districts not willing to hire more expensive Ph.D.s]

6. Given that funding were available, what incentives or special arrangements have been or could be implemented that would attract science and mathematics Ph.D.s to secondary school teaching?

Probes [Alternative certification process: what would it look like?]

[Enhanced compensation: salary, merit pay, summer fellowship, student loan repayment, relocation reimbursement, benefits, other]

[Placement in magnet schools or schools for gifted students]

[Ability to contribute to curriculum development]

[Ability to conduct research]

[Ability to job share]

[Availability of lab quality lab equipment or science kits]

[Special summer programs]

[Other]

7. Are any of these already available due to policies implemented by the state, the school district you are in, or your school?
8. Which of these could additionally be implemented at the high school level?
9. What arrangements have been or could be implemented by graduate schools that would facilitate the transition of science and mathematics Ph.D.s to secondary school teaching?

Probes [Availability of teacher preparation courses to Ph.D. candidates]

[Partnerships between universities and school systems that provided exposure of grad students to the K-12 environment]

[Availability of fellowships that gave students an opportunity to work in middle or high schools while in graduate school]

10. Are any of these currently available? Which might be implemented by colleges or universities in your state?
11. How willing are you to allow a reduced teaching load for Ph.D.s during a training period??
12. Are their funds available for teacher training for Ph.D.s?

NATIONAL RESEARCH COUNCIL

Office of Scientific and Engineering Personnel Center for Science, Mathematics, and Engineering Education

Attracting Science and Mathematics Ph.D.s to Secondary School Teaching

Interview Guide:

School Superintendents

Hello, this is [name of interviewer] with the National Research Council. I am working on a NRC project investigating the potential for attracting recent science and mathematics Ph.D.s to secondary school teaching or curriculum development positions. I hope you received the letter we sent you last week introducing you to this project.

As one step in this study, we are conducting a national survey of graduate students and recent Ph.D.s to assess their level of interest in such positions, the obstacles they face in taking them, and the incentives they would need to make the transition to these positions. As another step, we would like to obtain the views of high school principals, district superintendents and state education commissioners on the prospects for placing Ph.D.s in these positions.

1. Do you see recent Ph.D.s as a source of quality science and mathematics teachers for the high schools or middle schools in your school district?
2. Why or why not?
3. What do you see as the advantages of Ph.D.s as secondary school science and mathematics teachers?

Probes [Knowledge of subject at the forefront of the field]

[Commitment to subject]

[Ability to convey scientific methods to students]

[They would make fine teachers]

[Would provide an alternative career for them—would anyone say this?]

4. What do you see as the disadvantages of Ph.D.s as secondary school science and mathematics teachers?

Probes [Lack of teaching ability]

[Lack of commitment to secondary education]

[They're trained for and mainly interested in research]

[Unable to thrive in K-12 environment]

[Unable to relate to other faculty]

5. What are the obstacles to attracting and retaining science and mathematics Ph.D.s as secondary school teachers?

Probes [Lack of teacher certification]

[Need for further training in education]

[Lack of interest or commitment on the part of Ph.D.s in teaching]

[Would not be made welcome by other teachers]

[Salary/compensation levels are not sufficient to attract them]

[School districts not willing to hire more expensive Ph.D.s]

6. Given that funding were available, what incentives or special arrangements have been or could be implemented that would attract science and mathematics Ph.D.s to secondary school teaching?

Probes [Alternative certification process: what would it look like?]

[Enhanced compensation: salary, merit pay, summer fellowship, student loan repayment, relocation reimbursement, benefits, other]

[Placement in magnet schools or schools for gifted students]

[Ability to contribute to curriculum development]

[Ability to conduct research]

[Ability to job share]

[Availability of lab quality lab equipment or science kits]

[Special summer programs]

[Other]

7. Are any of these currently available through school district policies? Which of these could be implemented at the school district level?
8. What arrangements have been or could be implemented by graduate schools that would facilitate the transition of science and mathematics Ph.D.s to secondary school teaching?

Probes [Availability of teacher preparation courses to Ph.D. candidates]

[Partnerships between universities and school systems that provided exposure of grad students to the K-12 environment]

[Availability of fellowships that gave students an opportunity to work in middle or high schools while in graduate school]

9. Are any of these currently available? Which might be implemented by colleges or universities in your state?
10. How willing do you think schools would be to allow a reduced teaching load for Ph.D.s during a training period?
11. Are their funds available at the school district level for teacher training for Ph.D.s?
12. Are there other positions in which you would like to see Ph.D.s contribute?

NATIONAL RESEARCH COUNCIL

Office of Scientific and Engineering Personnel Center for Science, Mathematics, and Engineering Education

Attracting Science and Mathematics Ph.D.s to Secondary School Teaching

Interview Guide:

Chief State School Officers or State Policymakers

Hello, this is [name of interviewer] with the National Research Council. I am working on a NRC project investigating the potential for attracting recent science and mathematics Ph.D.s to secondary school teaching or curriculum development positions. I hope you received the letter we sent you last week introducing you to this project.

As one step in this study, we are conducting a national survey of graduate students and recent Ph.D.s to assess their level of interest in such positions, the obstacles they face in taking them, and the incentives they would need to make the transition to these positions. As another step, we would like to obtain the views of high school principals, district superintendents and state education commissioners on the prospects for placing Ph.D.s in these positions.

1. Do you see recent Ph.D.s as a source of quality science and mathematics teachers in high schools in your state?
2. Why or why not?
3. What do you see as the advantages of Ph.D.s as secondary school science and mathematics teachers?

Probes [Knowledge of subject at the forefront of the field]

[Commitment to subject]

[Ability to convey scientific methods to students]

[They would make fine teachers]

[Would provide an alternative career for them—would anyone say this?]

4. What do you see as the disadvantages of Ph.D.s as secondary school science and mathematics teachers?

Probes [Lack of teaching ability]

[Lack of commitment to secondary education]

[They're trained for and mainly interested in research]

[Unable to thrive in K-12 environment]

[Unable to relate to other faculty]

5. What are the obstacles to attracting and retaining science and mathematics Ph.D.s as secondary school teachers?

Probes [Lack of teacher certification]

[Need for further training in education]

[Lack of interest or commitment on the part of Ph.D.s in teaching]

[Would not be made welcome by other teachers]

[Salary/compensation levels are not sufficient to attract them]

[School districts not willing to hire more expensive Ph.D.s]

6. Given that funding were available, what incentives or special arrangements have been or could be implemented that would attract science and mathematics Ph.D.s to secondary school teaching?

Probes [Alternative certification process: What would it look like? Is something different needed?]

[Enhanced compensation: salary, merit pay, summer fellowship, student loan repayment, relocation reimbursement, benefits, other]

[Placement in magnet schools or schools for gifted students]

[Ability to contribute to curriculum development]

[Ability to conduct research]

[Ability to job share]

[Availability of lab quality lab equipment or science kits]

[Special summer programs]

[Other]

7. Are any of these currently available through state-level policies? Which of these could additionally be implemented at the state level?
8. What arrangements have been or could be implemented by graduate schools that would facilitate the transition of science and mathematics Ph.D.s to secondary school teaching?

Probes [Availability of teacher preparation courses to Ph.D. candidates]

[Partnerships between universities and school systems that provided exposure of grad students to the K-12 environment]

[Availability of fellowships that gave students an opportunity to work in middle or high schools while in graduate school]

9. Are any of these currently available? Which might be implemented by colleges or universities in your state?
10. Are their funds available at the state level for teacher training for Ph.D.s?
11. Are there other positions in which you would like to see Ph.D.s contribute?

NATIONAL RESEARCH COUNCIL

Office of Scientific and Engineering Personnel Center for Science, Mathematics, and Engineering Education

Attracting Science and Mathematics Ph.D.s to Secondary School Teaching

Interview Guide:

Graduate Deans

Hello, this is [name of interviewer] with the National Research Council. I am working on a NRC project investigating the potential for attracting recent science and mathematics Ph.D.s to secondary school teaching or curriculum development positions. I hope you received the letter we sent you last week introducing you to this project.

As one step in this study, we are conducting a national survey of graduate students and recent Ph.D.s to assess their level of interest in such positions, the obstacles they face in taking them, and the incentives they would need to make the transition to these positions. As another step, we would like to obtain the views of graduate deans regarding the prospects for training Ph.D.s for these positions.

1. Do you see recent Ph.D.s as a source of quality science and mathematics teachers in high schools [in your state]?
2. Why or why not?
3. What do you see as the advantages of Ph.D.s as secondary school science and mathematics teachers?

Probes [Knowledge of subject at the forefront of the field]

[Commitment to subject]

[Ability to convey scientific methods to students]

[They would make fine teachers]

[Would provide an alternative career for them—would anyone say this?]

4. what do you see as the disadvantages of Ph.D.s as secondary school science and mathematics teachers?

Probes [Lack of teaching ability]

[Lack of commitment to secondary education]

[They're trained for and mainly interested in research]

[Waste of research talent]

[Unable to thrive in K-12 environment]

[Unable to relate to other secondary school faculty]

5. What are the obstacles to attracting and retaining science and mathematics Ph.D.s as secondary school teachers?

Probes [Lack of teacher certification]

[Need for further training in education]

[Lack of interest or commitment on the part of Ph.D.s in teaching]

[Would not be made welcome by other teachers]

[Salary/compensation levels are not sufficient to attract them]

[School districts not willing to hire more expensive Ph.D.s]

6. What arrangements have been or could be implemented by graduate schools that would facilitate the transition of science and mathematics Ph.D.s to secondary school teaching?

Probes [Availability of teacher preparation courses to Ph.D. candidates]

[Partnerships between universities and school systems that provided exposure of grad students to the K-12 environment]

[Availability of fellowships that gave students an opportunity to work in middle or high schools while in graduate school]

7. Are any of these currently available? Which would you be willing to implement?
8. What incentives or special arrangements have been or could be implemented by school systems to attract science and mathematics Ph.D.s to secondary school teaching?

Probes [Alternative certification process: what would it look like?]

[Enhanced compensation: salary, merit pay, summer fellowship, student loan repayment, relocation reimbursement, benefits, other]

[Placement in magnet schools or schools for gifted students]

[Ability to contribute to curriculum development]

[Ability to conduct research]

[Ability to job share]

[Availability of lab quality lab equipment or science kits]

[Special summer programs]

[Other]

9. Are any of these currently available? Which of these could additionally be implemented by states, school districts, or schools?

Appendix D

Interviews with Ph.D.s in K-12 Science and Mathematics Education

To complement the survey of graduate students and recent Ph.D.s about their potential interest in K-12 education, we interviewed 18 Ph.D.s already working at the K-12 level to draw insights from their experiences about the opportunities for—and obstacles to—Ph.D.s in K-12 education careers.

These interviews were conducted by telephone during the month of August 1999. NRC staff compiled a short list of Ph.D.s working in teaching or curriculum development positions and these individuals were contacted for an interview. At the end of each interview, we asked the individual being interviewed if he or she could think of someone else who would be suitable for interviewing. This "snowball" sampling method yielded more than 32 names, of which 20 were available for interviews in the time allotted. Two of the 20 did not hold doctorates, so the total number of Ph.D.s interviewed was 18.

In interviews typically lasting 35 to 50 minutes, we asked these individuals to talk about their experiences in K-12 education. We asked them to comment on what attracted them to secondary school teaching or to curriculum development. We asked them whether colleagues or mentors were supportive when they decided to enter K-12 education, to describe any barriers they encountered in taking K-12 education positions, and how they handled teacher certification. We also asked for the effects of financial or family considerations on their career paths. Individuals interviewed were asked to comment on their current work environment, their relationships with other faculty and administrators, and whether they were actively engaged in research. Finally, we asked them whether they were happy with their choice of a career working in K-12 education, what advice they would give a Ph.D. considering such a career, and what the prospects for such a career would be at this time.

While the interviewer asked these specific questions, the interview was also conducted in such a manner as to encourage the interviewee to explore the reasons why they went into teaching and to discuss what were the most significant issues involving that career choice. Responses to specific questions are detailed below, followed by a section that summarizes the many ideas for improving the experience of Ph.D. teachers that were elicited from the interviews.

CHARACTERISTICS OF INDIVIDUALS INTERVIEWED

Out of a list of 32 individuals working in K-12 science or mathematics education, 20 were contacted and 18 were found to hold a doctorate. Respondents were located in all regions of the country. The characteristics of these 18 individuals are as follows:

- Of the 18 Ph.D.s interviewed, 14 had received their Ph.D. prior to beginning a career in K-12 education and four received the doctorate after beginning such a career.
- Eight had held postdoctoral fellowships and 10 had not.
- Sixteen were teaching or had taught in secondary schools (2 with extensive experience) and two were working in curriculum development. Of the 16 teachers, 10 were still teaching, five had left teaching, and one had left and returned to it later.
- Eight of the 18 Ph.D.s interviewed were teaching or had taught at private/independent schools and three of the eight private schools were religious in orientation.
- Seven of the 18 had taught at the college or university level prior to entering the secondary education environment.
- Five of the 16 teachers earned their teaching certificate prior to completing their Ph.D., seven earned certification after completing the Ph.D., and four (all of whom were teaching in private schools) did not obtain teacher certification.
- The two individuals working in curriculum development also did not acquire teacher certification.
- The Ph.D.s were in a wide range of fields: biology/physiology (six); earth sciences/geology (four); mathematics (three); physics (two); biochemistry (two); and computer education (one).
- Ten were male and 8 were females.

FOLLOWING A CAREER IN K-12 EDUCATION

1. What attracted you to secondary school teaching?

- A common theme, cited by 10 respondents, was that they had gone into teaching because they loved teaching, enjoyed working with children, and enjoyed helping children succeed.

- One respondent stated: "*Sometimes I feel like a bull fighter: if you stand right in front of the bull* [a.k.a. high school students] *you will get squished; but if you point them in the right direction, they will take off and go right past you and keep going.*" This was a very satisfying experience for him, to see the children learn how to learn.
- Respondents frequently had more than one reason for entering teaching, however, and there was a great variety in the responses:

1. One Ph.D. teacher, who had been a university administrator for 20 years, stated that "*Going into high school teaching meant he could go back to thinking almost full time about math and science. A chance to really learn the stuff that he thought he had learned a long time ago.*"
2. Several complained that they found research too isolating, and not suited to their characters because they were extroverted.
3. Two mentioned that they had physical injuries that precluded them from continuing in the physically demanding arena of research.
4. Three really didn't want to teach high school. Two felt it was necessary for career development since they ultimately wanted to return to college level teaching in curriculum development. One said it permitted him to do research in the summer months in his area of specialization.
5. Several people referred to the feeling that they were giving something back to the community. One stated: "*I'm not so excited about creating a billion scientists, but science literacy is something I really believe in, as an equalizer or way for people to take care of themselves, to learn about the world, and to make sure they are not taken advantage of, to preserve the environment…those are all important things that you can address best with K-12 education.*"

2. *Were your colleagues or mentors supportive when you made your decision to enter into secondary school teaching? Did you meet with approval or disapproval?*

- There were three common responses to this question:

1. Eight reported that mentors and colleagues were generally supportive. The geologists who were turning to K-12 careers during a downturn in the oil industry found others to be especially supportive. One respondent indicated that her mentor was supportive of her decision.
2. Six reported that the decision was met by "total disbelief," mostly from colleagues at high profile research institutions. Several respondents reported that others told them they were wasting their education. One teacher claimed that a professor from her alma mater told her this year that "*we are not training our students here to be high school teachers. We are investing $100,000 per student… they need to return that investment to the scientific community.*" He did not thinking high school teaching was a return on that

investment. Another stated that colleagues "*thought her career was going well in research/ academia, so they didn't understand why she wanted to stoop to high school teaching.*"

3. Four reported no reaction to their decision.

3. *Did you encounter any barriers to taking a secondary school teaching position? How did you overcome these barriers?*

- Eight of the 18 Ph.D.s interviewed—six teachers and two in curriculum development—said there were no barriers whatsoever when they sought their K-12 education positions. Three of these six teachers were teaching in private schools. In both private and public institutions, the school administration was enthusiastic about their Ph.D.s, (almost like bragging rights) and the parents were very happy to have such qualified people in their children's classrooms.
- Eight teachers (three private, five public), however, stated that the teacher certification process was a real barrier to getting started. [See question #6 for more comments on this issue.]
- Five teachers (two private, three public) said they had a really hard time convincing the administration on their first job interviews that they were serious about their desire to teach. One stated: "*Most people don't trust Ph.D.s...they think they won't be on the mark; that they won't be able to relate to other people, that they won't be a team member. Also a lot of people are afraid of being upstaged.*" Another thing he encountered on interviews was "*why in the world would you take a Ph.D. and teach high school...you must be a failure.*" Several of them were hired only because the school district was in a panic to find someone that year. Subsequent jobs were much easier to land, once they had proven they were committed to this career path.
- One stated that he couldn't teach in private schools in the South because of his religious persuasion, since most of the private schools in the South were of a religious orientation.
- One person in curriculum development was worried about her career path because she had never taught in secondary science classrooms, but was responsible for organizing the volunteer program at a nearby university and preparing science kits. She wondered just how far she could go without the teaching experience.

4. *To what extent did financial considerations have an effect on your decision to enter into secondary school teaching?*

- A striking finding from this question is the number of respondents who said they couldn't have chosen the teaching profession unless they had sufficient family income so they didn't have to worry about their salaries: seven said they had their spouses' financial support, and five had saved funds (retirement, stocks) from a previous career. One person stated that it was "*the risk of whether she would like it or not that was more of a consideration than the financial considerations.*" Another stated that she felt more secure in secondary education than in college or university settings because she was not dependent on grants, etc. Another stated it was the working conditions (long hours, lack of laboratory support) and lack of prestige, not the lack of pay, that was driving teachers away.

- Except for respondents from Washington and New York State, in most cases low salaries were an area of concern. In Washington, the Ph.D.s are paid by the state, not the school district, so they can be paid more without dipping into the school districts' resources. In New York, one teacher with 20 years experience and several published textbooks said he was making more than $90,000 for 9-months work, hence he was making more as a teacher than a researcher would be making.
- Two said that salary wasn't a consideration, they were just glad to have a job.

5. *To what extent did family considerations have an effect on your decision to enter into secondary school teaching?*

- There were three general responses to this question:

1. Four respondents didn't have children, or the children were grown up, so family considerations were not an issue. One mentioned if he had children, he wouldn't have been able to stay in the teaching profession because he would have needed more money.
2. Eight said that having more time with their children (mostly summers, but also after school hours) was a benefit, but not the only reason they chose this profession. One woman stated that "*doing something you like becomes more important when you have to be separated from your child to do it.*" Another said 99.9 percent of the reason he went into teaching, was "*so that I could spend more time with my family. My research colleagues were married to their jobs.*" Another woman stated that "*the women scientists that were successful were not people that she wanted to emulate.*" She was told her "*best strategy as a female scientist is to marry someone rich so you won't have to worry about paying for* [dawn-night] *childcare.*"
3. Nine said that they were faced with geographic restrictions due to family issues (spouse's job or custody issues) so they were constrained to find a job that they would like within a narrow geographic focus.

- An additional two people mentioned they had specifically chosen to teach in private rather than public schools so that they could get a tuition break for their children at that school.

6. *How did you obtain teacher certification? Was this the typical way teachers are certified in your state or was this an alternative certification process? What alternative certification processes are available in your state? Did your state (and/or the State Department of Education) provide you any assistance or pose any obstacles in getting you certified? If yes, how.*

- Five of the 16 Ph.D. teachers received their teaching certificates while undergraduate students. Two took education classes for the certificate while teaching at a college. Three took summer/evening classes while teaching at an independent school or teaching at a local community college. One was able to convince the university to let her do the certification via independent study, since she was faculty there and was teaching in two of the three required courses. One person took off an entire year to get a master's in Education. He estimated it cost him over $20,000 (tuition, lost income) to go this route. Six did not get certification (four taught at independent schools, two were in program development).
- Ten people complained about the teacher certification process, noting most frequently that the teacher education classes were useless (eight), or near to useless (two). One noted a course that included instruction on how to use a transparency machine.
- At the same time several of the teachers with certification felt that it was a good idea to at least learn how children think and learn, or to do some supervised student teaching where they got constructive feedback. The repeating theme was that just knowing the material was not sufficient to be able to teach it well. One person stated "*If I went in there with what I thought was teaching, no way would it have worked in middle schools.*" Another private school teacher stated "*it is a big mistake that private schools don't require certification.*" His [teacher education] program was abysmal, but it did make him think about teaching and learning about alternate ways to present the material.
- Many felt that an accelerated, or boot camp, approach to teacher certification would have been sufficient, while others said 6–10 weeks was not sufficient. One respondent stated: "*it is a feasible option to speed up the teacher certification process.* [It] *may not be ideal, but you can't expect someone who spent years and years getting a Ph.D. to be willing to spend three semesters to get their teaching certificate.*"
- Some mentioned that the cost was a substantial barrier in those states that didn't have an accelerated program, as there was both tuition and lost salary to consider.
- Aid from the states was non-existent or very limited. Emergency or probationary certificates were the only things mentioned as possible aid from the state.

WORKING IN K-12 EDUCATION CAREERS

1. Would you please describe the school in which you are currently teaching?

- Eight were teaching or had taught in private schools, ten were in public schools, and two were in curriculum development. The private schools were either college-preparatory (five) or religious (three) in nature, and were typically less than 500 students in the secondary education division. The public schools ranged in size from 500-2000 students.
- The number of courses taught in either private or public schools per day did not seem to be significantly different between the two school types (four to five courses each), but the class

sizes were definitely smaller in the private (mostly less than 22 students/class) than in the public (high 20's) schools. Teachers at private schools were not required to coach or do cafeteria duty.

- About the same number of teachers (four private vs. three public) mentioned teaching honors classes at their schools, but only the public school teachers mentioned teaching remedial courses (two). One Math Ph.D. said he preferred teaching the remedial classes.
- Seven had left teaching in high schools, but all felt they were still teaching, albeit in a nontraditional environment: three taught educational workshops at research institutions, one is now teaching at a community college, one is preparing science kits for statewide distribution, one is coordinating teacher education program at major research university, and one has returned to teaching after doing well in the stock market.

2. *What are your relationships with other faculty and with the administration like? Are there other Ph.D.s teaching at the school?*

- Eleven interviewees felt that their holding a Ph.D. was not an issue in the high school or curriculum development setting, though four mentioned there was some initial resistance or caution amongst the colleagues, until they had proven themselves as good teachers. One stated: "*If anything,* [the Ph.D.] *might be a hindrance. You know all this stuff, and if you share it, you may come across as pontificating, or being bossy. Got to keep your mouth shut and do it.*" Another noted that personality does play a big part in this issue, and that "*a Ph.D. that goes into teaching is not typically arrogant.*"
- Five said school administrators loved to brag about the number of Ph.D.s they had on staff. In one case, the respondent was the only Ph.D. among 150 teachers. In public schools, respondents were frequently the only Ph.D. science teachers on staff, although there would usually be some Ed. D. administrators. At private schools, other Ph.D. colleagues would be more likely.

3. *Are you still engaged in research?*

- The majority of people interviewed said they were not doing traditional research-like activities (11), while four people said they were maintaining their research interests during the summer. One person deliberately chose teaching so that he could do research in the summers. He stated: "*There are a lot of high school teachers that live and breathe high school 12 months of the year, but I am just not interested in that. Frankly, I spend 9 months of the year with 14- and 15-year olds, and I am not interested in spending my summer months with them too*". This same person later stated, "*If you are not looking at maintaining the research end, they* [Ph.D. teachers] *would have to reconcile why they want to do high school teaching with the time and effort to get the Ph.D. If they can't reconcile this, they might have to face problems with dissatisfaction later.*"

- One had gone back to full-time research at a research institute, but was now also producing teacher education workshops there.
- Instead of pursuing research, many respondents were engaged in other professional activities. Six (four at private schools) were actively involved in producing teacher workshops. Three were writing textbooks (two were at private schools). In addition, individuals from private schools could be found to be reviewing grants for NSF or other funding associations (three), adapting computer programs to the high school setting (two), or doing private consulting (two).
- Several people said they believed research was too isolating for them—so much so that they couldn't continue in this field.

4. *Are you happy with your choice to become a secondary school teacher?*

- The majority of the people interviewed appeared to be happy or very happy with their choice. They stated that they really enjoyed watching students learn (nine) or teaching a particular subject (five). One specifically stated, "*When I teach, the greatest satisfaction is seeing the creativity of the kids, the way they approach problem solving, it is the 'ah-hahs'; they thought of something that I didn't even think about…that is fun.*" Two also mentioned they enjoyed forming long-term relationships with the kids. Five of the teachers in the private school setting volunteered that they were "very happy" with their choice. Public school teachers never said they were unhappy, but only one made the statement about being "very happy." Another teacher noted that teaching was a creative process, that one never does the same thing twice. Another stated, "*I believe* [public school education] *is where we should be putting our very best people. The kids with the greatest needs need the greatest teachers.*"
- There was very little overlap in the responses to the question about sources of frustration with teaching. Some of the statements made were:

(1) Bureaucracy and politics, and culture of adversity between administration and the teachers made the work environment frustrating (eight). One teacher said, "*The worst part was being treated like an employee, in the category of a servant…no independence… no prestige.*" Another (who had been in university administration for 20 years) said that he was dismayed by the number of high school administrators who were more concerned with "*getting the trains to run on time*" than trying new approaches to old problems.

Two teachers, both in public schools, said discipline is a problem.

1. The poor socioeconomic condition of the students was a source of concern. One person was quite depressed about this and said: "*Education is an insurmountable problem. There is just no way… these kids are going to be able to learn this, they are not going to have good lives.*" Because of the turmoil in the students' lives, the teacher had to be flexible and be able to adjust to surrounding situations rapidly.

2. Other problem areas included: (1) lack of high standards for the students or teachers, (2) lack of prestige, (3) rigid schedule with long hours, and (4) lack of help with laboratory classes.

It was somewhat surprising that only one person complained about lack of equipment or poor facilities (at a public school).

PROSPECTS FOR RECENT PH.D.S ENTERING CAREERS IN K-12 EDUCATION

1. What advice would you give to a science/math Ph.D. who is considering secondary school teaching?

- One overwhelming response was that it was essential for the candidate to get some high school teaching experience, either as a volunteer or as part of a certification process (13).
- Another piece of advice was for potential K-12 teachers to assess their personality, to determine if they are the types of persons who like working with people, in a cooperative setting, which is very different from the research environment (three).
- Eight respondents suggested that if graduate students were already past the master's level, and close to getting the Ph.D. that they should finish the Ph.D.—it would open some doors for you in the future. However, the Ph.D. didn't always make you a better teacher. If you knew before you got the master's that you wanted to teach rather than do research, one person suggested you stop at the master's level and jump right into teaching. One noted the people "*that are leaving science are not the ones that were bad scientists. Too many people are staying around because they don't have a clue as to what else they can do.*"
- In the same vein, another (who had returned to full time research) noted: "*I am not afraid of switching careers…I will try teaching again if I get bored with the work at the* [research] *institute. It is really hard to change careers the first time, but after that it is no big deal.*" One respondent noted there was a group of graduate students at a major research university who felt like they were failures in the eyes of their peers. The students "*found research is cold and impersonal, but didn't know how to break through without starving to death. I just said do it, have some faith in themselves. After all, they have half the battle taken care of: they know the content, they have time to focus on their people skills*".
- Several recommended that all K-12 teachers should get certified, even if they plan to work at a private school. Having the certificate would give them more options and there is some value in the psychology/ pedagogy courses.
- Two said not to do it unless you were sure it was a viable option, not to pursue teaching just because you didn't like research. You had to love teaching to put up with the low pay and long hours.

- When asked whether he was glad he had finished the Ph.D., one teacher said, "*I was really glad to have done the intense work, so I could concentrate on the teaching and model being a scientist. You can afford to have fun!*"
- Another piece of advice to future Ph.D. teachers was: "*Don't try to get a full-time job right away; having three different preps in the first year is overwhelming, and as a Ph.D., you tend to be too thorough... and try to avoid teaching at the 'bad' schools where you also have discipline problems to juggle. Take a step back and give yourself a chance to excel*".

2. *What do you think about the future of secondary school science teaching? What future do you envision for science and mathematics doctorates entering into secondary school teaching?*

- Twelve believed that the future for science teaching was "good" to "very good," especially in light of the new science curriculum reform movement. They felt more teachers were needed in this area, and they had to be better trained.
- They provided several recommendations that they thought would make the experience even better. Specifically, the teachers recommended that the secondary school system:

- Have better equipment in the labs;
- Have more infrastructure support (classrooms, facilities, etc.);
- Solve discipline problems;
- Avoid using fads in teaching style;
- Have smaller class sizes;
- Improve the quality of the teachers, especially those who "taught to the test" just to get pay raises; and
- Have a mix in the schools, of those with Ph.D.s and those without. Each has strengths to bring to the classroom, and can complement each other.

DISCUSSION

The majority of people interviewed were content or very happy with choosing a career in K-12 education. A frequent comment was that they had found research isolating and that they were much happier in a job that required interaction with other people. However, they also emphasized that you had to love teaching to put up with the long hours and low pay.

Teaching appeared to be a viable alternative to many faced with the two-career dilemma. Many of the respondents had made the decision to go into teaching due to family or geographic restrictions related to the spouse's job. The flip side of this coin, as revealed by subsequent probing, is that many are dependent on their spouses' income for a sense of financial security. While interviewees they were not asked about retirement benefits, and others were actually teaching now while in "retirement" from previous jobs. Teaching is a "luxury" available to them because of these other sources of income.

The respondents were positive about the future for Ph.D. science teachers, seeing it as a growth field, but were also concerned by the lack of quality people in the "pipeline." Again, the issue of bright people finding "easier" jobs for more pay was mentioned.

The respondents asserted that the Ph.D. was a valuable tool in their teaching experience. The Ph.D. teachers claimed to be more flexible in their teaching styles. That is, they were not dependent on the textbook for exercises and were able to create more challenging environments in the classroom because of the variety of ways they could present the material. These Ph.D. teachers seemed ideally suited to develop laboratory exercises for the new problem-based, inquiry learning methodology that is currently in vogue.

If a Ph.D. candidate knows that s/he is seriously interested in teaching that person should volunteer, substitute teach, or find some other way to get into the classroom during graduate school so as to make an informed decision. If, prior to getting the master's degree, the graduate students realize they disliked research and really want to teach, many suggested to stop at the master's and jump into teaching right away. For individuals who already have the master's degree, however, interviewees recommended that they complete the Ph.D. because it would help open some doors in the future.

Teacher certification, as it exists today in many states, is a major barrier to Ph.D.s considering public middle or high school teaching careers.

SUGGESTIONS FOR FACILITATING K-12 EDUCATION CAREERS FOR PH.D.S

The following ideas emerged from the interviews as ways to encourage Ph.D.s to go into secondary education:

1. There is a need to involve the Ph.D. teachers in problem-based inquiry learning programs. Ph.D.s may excel as teachers in such programs, because they know the content and can concentrate on the delivery and interpretation of the data with the students. Frequently, teachers without Ph.D.s are afraid to use this style of teaching because they are afraid to tell their students "*I don't know*" when they get unexpected results. This is the nature of science: lots of unexpected results. Ph.D.s have significant experience with dealing with unexpected results and formulating new hypotheses. Ph.D.s interested in secondary education teaching could volunteer to help science teachers during these exercises, providing a knowledgeable resource and, in return, getting experience in the classroom.
2. Encourage Ph.D. teachers and research laboratories to develop long-term relationships. Ph.D.s would return every summer to research laboratories for 10+ weeks to do research. These are highly qualified people who could be very focused on a particular component of a research project. They would be likely to be of more value than a high school, college or graduate student, and they would be more likely to return in subsequent summers. In addition, laboratory scientists with whom the Ph.D. teachers work during the summer could

be a potential resource during the school year for student mentoring students or providing surplus laboratory supplies.

3. There is a need for more opportunities like the NSF GK-12 fellowship program that supports graduate students who want to improve their teaching skills while finishing their Ph.D.s.
4. Develop a "science teacher aid" career path within school systems. These aides could help set up laboratory experiments and provide an extra pair of knowledgeable hands during the class. This would provide an alternative to other technical careers such as working as a laboratory technician in the biotechnology research. Such positions would probably only require an associate's degree to be qualified and aides could potentially be hired at a relativity low cost to the school districts.
5. Improve the quality of the teacher education classes required for certification.
6. As a corollary to item 5, an accelerated certification process should be available for Ph.D.s, preferably subsidized so that they could afford to do it during a summer or semester term prior to completing their Ph.D. and having to make a career decision.
7. States should be encouraged to take responsibility for teacher salaries, especially Ph.D.s, rather than individual school districts. This would equalize the quality of teachers across the state, and would remove any barriers to hiring a Ph.D. within a school district. Washington State has such a plan.
8. Recommend that petroleum and aerospace industries look into supporting their Ph.D. employees who want to get into secondary education teaching. Some sort of subsidized program could be offered during periods of downturn in the industry cycle. This would keep the Ph.D. teachers in the industry programs (usually the ones giving short courses, etc.) in the geographic area, but would provide a resource that can be rehired when the economy booms again.
9. One Ph.D. (who used to be vice president of a major university and is now a math teacher in an independent school) would really like to see workshops offered only for Ph.D.s who were teaching science at the high school level. He wanted less concentration on the material, and more networking and discussion of ways of approaching the material and effective communication with the students.

Appendix E

Interviews with Administrators

We also interviewed local and state school administrators and higher education administrators to obtain their perspectives on the opportunities for Ph.D.s in K-12 education and how they believed administrative structures might adjust to facilitate the entry of Ph.D.s into K-12 careers. Interviews were conducted with high school principals (regular public high schools and science and technology magnet schools), school district superintendents, and chief state school officers to obtain information about opportunities for and concerns about employment of Ph.D.s in secondary school teaching. Interviews were also conducted with graduate school deans to ascertain the kinds of programmatic changes in graduate education that might help to prepare Ph.D.s for careers teaching in secondary schools.

HIGH SCHOOL PRINCIPALS

Two principals of public high schools were interviewed. Both respondents indicated they did not see science as mathematics Ph.D.s as a source of quality teachers for their high school or in their district. Both respondents believed "teacher preparation" and ability to deal with youth was more important than holding a Ph.D. Both respondents also indicated that salary was an obstacle to attracting Ph.D.s to teaching. Specific responses to interview questions follow.

Response to Interview Questions (N = 2)

1. *Do you see recent Ph.D.s as a source of quality science and mathematics teachers in your high school or other high schools in your school district?*

- Both said No.

2. *Why or why not?*

   - Teachers need to be a quality educator.
   - Ph.D.s don't have the ability to deliver information to students in a way that students are able to use it.
   - Ph.D.s need to be recruited and high schools don't currently need to recruit teachers.
   - Don't think Ph.D.s offer a "leg up" in terms of other teachers.
   - Don't need Ph.D.s to teach high school.

3. *What do you see as the advantages of Ph.D.s as secondary school science and mathematics teachers?*

   - If teacher trained, can impart knowledge.
   - Enhance school's standing with Middle States evaluation.

- Knowledge of subject matter.

4. *What do you see as the disadvantages of Ph.D.s as secondary school science and mathematics teachers?*

   - Alternative route to certification is not the way to go.
   - Subject matter knowledge can overspecialize people and impact their ability to teach at the secondary school level.

5. *What are the obstacles to attracting and retaining science and mathematics Ph.D.s as secondary school teachers?*

Number of Mentions	Item Mentioned
2	Salary

Other items mentioned once each:

- Ph.D.s want to do research and be published (can't do this in school environment).
- No resources for research.
- Academics is a socially safe environment while high schools are not.
- Ph.D.s don't usually belong in a high school classroom.

6. *Given that funding were available, what incentives or special arrangements have been or could be implemented that would attract science and mathematics Ph.D.s to secondary school teaching?*

   - Paid training period to learn teaching strategies and classroom management.
   - Use Ph.D.s to teach advanced placement classes.
   - Use good teachers to teach the worst kids (to get them excited about the subject area).

7. *Are any of these already available due to policies implemented by the state, the school district you are in, or your school?*

   - Training period to learn teaching strategies and classroom management (mentioned in #6 above) is currently required. New teachers are required to attend summer workshops, but are not paid for this. (N = 1)

8. *Which of these could additionally be implemented at the high school level?*

One respondent mentioned each of the following:

- Incentives beyond salary.
- Class size and teacher unions are stumbling blocks.

- Salaries need to attract young people to teaching; review the salary steps and match them to industry.

9. *What arrangements have been or could be implemented by graduate schools that would facilitate the transition of science and mathematics Ph.D.s to secondary school teaching?*

    - Provide teacher preparation (courses).
    - Provide classroom experience and orientation to schools through teaching.

10. *Are any of these currently available? Which might be implemented by colleges or universities in your state?*

    - Would be willing to discuss.

11. *How willing are you to allow a reduced teaching load for Ph.D.s during a training period?*

    - Yes, but has financial implications for district.
    - No, because of teacher unions.

12. *Are their funds available for teacher training for Ph.D.s?*

    - State handles this.
    - Not from individual districts, maybe from state.
    - District handles professional development activities for new teachers and offers professional.
    - Development three times a year for all teachers.

PRINCIPALS OF SCIENCE AND TECHNOLOGY (MAGNET) HIGH SCHOOLS

Three principals of specialized science and technology high schools (magnet schools) were interviewed. All three believe that the experience of Ph.D.s with scientific inquiry, experiments, and scientific enterprise are an advantage. However, all three interviewees indicated Ph.D.s needed to be exposed to secondary education, with time working in a school, in order for this to become a viable career option. All three believe a reduced teaching load during a training period would be beneficial. Two respondents mentioned money/salaries as obstacles.

Response to Interview Questions (N = 3)

1. *Do you see recent Ph.D.s as a source of quality science and mathematics teachers in your high school or other high schools in your school district?*

- Two – Maybe
- One - Yes

2. *Why or why not?*

Why:

- Attitude and energy.
- Content background.
- Can be a good resource.

Why Not:

- Been in academic world too long.
- Only deal with peers/colleagues.
- Speak own language.
- Must be flexible.
- Must make things relevant, understandable, and comprehensive for students.
- Must be sensitive to student needs.
- Have career goals other than secondary school teaching.
- Must have human relations qualities.
- Must have background in education.

3. *What do you see as the advantages of Ph.D.s as secondary school science and mathematics teachers?*

Number of Mentions	Item Mentioned
3	Experience in scientific inquiry, experiments, and scientific enterprise
2	Depth of knowledge
2	Broad view/scope of discipline
2	Knowledge base/big picture versatile
2	Lend credibility

Also mentioned were:

- Can institute national standards in science/math.
- Resourceful in getting school grants and additional resources.
- Subject matter perspective.
- Very focused.

4. *What do you see as the disadvantages of Ph.D.s as secondary school science and mathematics teachers?*

Number of Mentions	Item Mentioned
2	Need level of support (handholding from principals and provide them with "drive")

Other items mentioned by respondents included:

- They loose sight of the level kids can achieve.
- They are overspecialized when a generalist is needed.
- Pedagogical difference between high school and college.
- Financial needs.
- Goal is to publish, work in university or private industry, not teach.
- Have to love kids and want to be in their world.
- If goal is elsewhere, will only teach for a short time (2 to 3 years).
- In urban districts, the cost of housing is very high.

5. *What are the obstacles to attracting and retaining science and mathematics Ph.D.s as secondary school teachers?*

Number of Mentions	Item Mentioned
2	Money
2	Science chairs and principals may feel threatened; teachers may not see benefit of Ph.D.

Other items mentioned included:

- You teach kids, not a subject.
- Alternative certification needs to be augmented with mentoring.
- Schools are limited in resources and remain status quo; can't get resources or innovations.
- Often new teachers are given worst teaching assignments.

6. *Given that funding were available, what incentives or special arrangements have been or could be implemented that would attract science and mathematics Ph.D.s to secondary school teaching?*

Number of Mentions	Item Mentioned
2	Resources, such as merit pay, signing bonus, subsidized salaries

Other items mentioned included:

- Create research opportunities through funds.
- Develop relationship between education and science and math departments in universities.
- View teaching as a broader vision.
- Include secondary education as a career option.
- Give graduate students opportunities to visit schools.

- Have salaries sufficient enough so that teachers can travel and renew each summer without financial worries.

7. *Are any of these already available due to policies implemented by the state, the school district you are in, or your school?*

    - Need support of school board.
    - Need legislative push to do so (with science and math and education departments in universities working together).
    - Not available in state.
    - Doesn't think options are available in state.

8. *Which of these could additionally be implemented at the high school level?*

    - District may be interested; state probably believes they have enough teachers.

9. *What arrangements have been or could be implemented by graduate schools that would facilitate the transition of science and mathematics Ph.D.s to secondary school teaching?*

Number of Mentions	**Item Mentioned**
2	Require field experience for teaching, including adult evening classes or secondary school

Other mentions included:

- Link education department with science and math departments.
- Require students to do classroom observation.
- Require students to do a classroom internship.
- Help students with tutoring.
- Help teachers with staff development to inspire teachers.

10. *Are any of these currently available? Which might be implemented by colleges or universities in your state?*

    - Doesn't think any of these are being done.
    - Have made efforts to make connections with individuals within universities in certain departments (such as environmental education).
    - Found instructors at colleges are rigid; students cannot fast track into classroom teaching.

11. How willing are you to allow a reduced teaching load for Ph.D.s during a training period?

All three responded they would be willing to allow a reduced teaching load.
Qualifications made to the statement:

- If you do it for Ph.D.s you must do it for all rookie teachers.
- Had a grant for this, but funding ran out (was specific to science and math teachers). Teachers met with college professor and were mentored.

12. Are their funds available for teacher training for Ph.D.s?

- One person said it was up to the individual seeking certification; another said the school district would pick up a portion of the costs and grants were available for this. The third didn't comment on this issue.

Additional comments:

- When asked for additional information, respondents offered three comments: Ph.D.s need to know teaching is an option for them; people of color are needed as role models in the math and science arenas, especially teaching; and there needs to be small research projects funded at the high school level for students to do independent study.

SCHOOL DISTRICT SUPERINTENDENTS

Response to Interview Questions (N = 3)

1. Do you see recent Ph.D.s as a source of quality science and mathematics teachers for the high schools or middle schools in your school district?

- Two "maybe"
- One yes

2. Why or why not?

Why:

- State assessments require higher levels of content knowledge; important for both kids and in-service to teachers.
- Bring research knowledge.
- Bring subject matter knowledge.

Why Not:

- Not a realistic source.
- Want higher salaries than districts can pay.
- Don't need their level of expertise.
- No practical or theoretical experience for lesson activities in classroom.
- No education training.

3. *What do you see as the advantages of Ph.D.s as secondary school science and mathematics teachers?*

Number of Mentions	**Item Mentioned**
2	Body of knowledge/knowledge of subject

Also mentioned:

- Can use base of research.
- Can move beyond subject matter.

4. *What do you see as the disadvantages of Ph.D.s as secondary school science and mathematics teachers?*

    - Specialization when need a breadth of knowledge.
    - Pedagogy and teaching experience lacking.
    - Ability to design interactive lessons not present.
    - Student lessons have to model real world.

5. *What are the obstacles to attracting and retaining science and mathematics Ph.D.s as secondary school teachers?*

Number of Mentions	**Item Mentioned**
3	Salary/money

Also mentioned:

- The need for the individual to be a generalist.
- Lack of ability to focus on specialist area of interest.
- Ph.D.s are interested in university or research institute more than secondary school environment.
- Working conditions (no time for reflection or planning)
- Some areas of need (rural) may not have amenities, plus people are generally less educated than Ph.D.

6. *Given that funding were available, what incentives or special arrangements have been or could be implemented that would attract science and mathematics Ph.D.s to secondary school teaching? (Note: Items with "*" have already been implemented or are available at some level in the respondent's state.)*

- Partnerships with businesses or universities so Ph.D.s can be involved in schools and still pursue specialized area of interest.
- Magnet schools, but you don't want to narrow kids' exposure or choices too early.
- Use Ph.D.s as mentors to students.*
- Pair them with someone who understands the realities of public schools and have them do curriculum development.
- Provide release time for training* and for research.
- Adequate salaries (legislature should increase salaries for Ph.D.s so they remain competitive with business).
- Summer institutes to prepare Ph.D.s for teaching.
- Lower teacher loads and improve workplace environment.

7. *Are any of these currently available through school district policies? Which of these could be implemented at the school district level?*

The release time for teachers is part of a NSF grant program and is only available in certain districts. Other items mentioned included:

- Some of these things would not become school board policy.
- Some universities might be willing to work with alternative certification.
- Schools have to want the Ph.D.s as much as the Ph.D.s want the school.
- If funding were available, district would be likely to implement these ideas.

8. *What arrangements have been or could be implemented by graduate schools that would facilitate the transition of science and mathematics Ph.D.s to secondary school teaching?*

Number of Mentions	Item Mentioned
3	Ph.D.s should spend time in the schools

The specific comments for this mention included offering scholarships or fellowships to graduate students to spend time in the schools; requiring Ph.D.s to perform internships in the schools; and students should work with schools and real students.

Other items mentioned in response to this question included:

- Fund students who want to look at public education (it's looked down on).
- Have Ph.D.s obtain basic education training while in graduate school.
- Become resources (especially with technology) to teachers in the schools.

9. *Are any of these currently available? Which might be implemented by colleges or universities in your state?*

- Respondents were vague about this. They said they didn't know why it wouldn't be done; assume it would be done if resources were there; and stated that universities said they would do these things, but didn't know if they would follow through.

10. *How willing do you think schools would be to allow a reduced teaching load for Ph.D.s during a training period?*

Number of Mentions	Item Mentioned
3	It's a finance/funding issue

Other responses included:

- Depends on the funds available.
- Teacher unions are an issue.
- Only accepted if it reduces the load all the way around (across teachers, not specific to Ph.D.s).
- Not sure other teachers would see advantage of having Ph.D.s in classroom (particularly if this was done).

11. *Are their funds available at the school district level for teacher training for Ph.D.s?*

- Two responded "yes."
- One responded "no."
- For the "yes" responses, comments included: not specific to Ph.D.s, funding may be available but not budgeted, and state has program for individuals willing to work in specific poor, rural counties in state.
- Mentors to other students, to keep them up-to-date and excited about subject content.
- Serve as in-service specialists, training consultants, and writers for curriculum.

12. *Are there other positions in which you would like to see Ph.D.s contribute?*

Responses included:

- Work in instructional design, using instructional technology.

CHIEF STATE SCHOOL OFFICERS OR STATE POLICYMAKERS

Response to Interview Questions (N = 3)

1. *Do you see recent Ph.D.s as a source of quality science and mathematics teachers in high schools in your state?*

- Two – Yes
- One – No

2. *Why or why not?*

Why:

- Ph.D.s can help students achieve rigorous core curriculum content standards/state student testing.
- Have content knowledge base and ability to find appropriate information.
- Magnet schools for science and math.
- Enhance student capacity for students achieving at higher levels in science and math.

Why Not:

- They go on to higher level jobs in education.
- Must be good at pedagogy.

3. *What do you see as the advantages of Ph.D.s as secondary school science and mathematics teachers?*

   - Ph.D.s will bring back respect and professionalism to teachers.
   - Enliven other staff.
   - Content specialists.
   - Advanced level of training and exposure to content knowledge.
   - Additional source from which quality teachers can be found.

4. *What do you see as the disadvantages of Ph.D.s as secondary school science and mathematics teachers?*

Number of Mentions	Item Mentioned
2	Other teachers may see as a threat

Other disadvantages mentioned by interviewees included:

- Higher level of expectation.
- More demands.
- Won't put up with things.
- Principals are threatened by Ph.D.s.
- System is too difficult.
- Ph.D.s not prepared for problems in classroom.
- No intellectual reflection.
- May not see job as teaching kids.
- Need to know child development, child learning, and teaching.

5. *What are the obstacles to attracting and retaining science and mathematics Ph.D.s as secondary school teachers?*

Number of Mentions	Item Mentioned
3	Salary
3	Certification process (not all districts use alternative certification, and the ease with which can teach without certificate does not provide incentive to get certificate)

Other items mentioned as obstacles included:

- No monetary rewards for advanced education.
- Some personnel (in schools) may feel threatened.
- Working conditions in schools (facilities and equipment).
- No support systems for teachers.
- Disruptive classrooms.
- Little respect.
- "Too many kids and too many preps."

6. *Given that funding were available, what incentives or special arrangements have been or could be implemented that would attract science and mathematics Ph.D.s to secondary school teaching?*

- Time for reflection.
- Time for research.
- Academic freedom from curriculum.
- Opportunities to mentor and network.
- More funds to pay for alternate certification.
- Job sharing with private industry or governmental agency.
- Better salary.
- Better working conditions.
- Fewer bureaucratic hoops to go through.

7. *Are any of these currently available through state-level policies? Which of these could additionally be implemented at the state level?*

One state has several programs for teachers. In one, teachers worked with university professors on a research project to learn scientific inquiry and statistics. The respondent indicated many teachers dropped out because this effort was too demanding. The same state also offers teachers an opportunity to join regional collaboratives. These collaboratives usually focus on a research project each year. One collaborative is helping teachers obtain additional education (but not a hard science and math Ph.D.).

Another state is using grant funds to focus on recruiting teachers, particularly in the eastern seaboard states. Recruitment from colleges and universities includes individuals who would be alternate certification candidates, such as science and math Ph.D.s. Finally, the individual who suggested job sharing indicated this would require state action in order for it to occur.

8. *What arrangements have been or could be implemented by graduate schools that would facilitate the transition of science and mathematics Ph.D.s to secondary school teaching?*

Number of Mentions	Item Mentioned
2	Send Ph.D.s into the classroom

Other possibilities for graduate schools offered by respondents included:

- Ph.D.s mentoring teachers and providing time for reflection.
- Students participating in a computerized school district job database (to explore job possibilities with the school districts).
- Linking with professional associations, such as school administrators, school principals.
- Coursework in pedagogy.
- Weighting outreach to public schools to be important as research and teaching for tenure.

9. *Are any of these currently available? Which might be implemented by colleges or universities in your state?*

Respondents were undecided as to whether universities would be willing to implement any of these options. One respondent commented it was "difficult" making linkages with the "pure sciences" within universities. Another respondent indicated some universities would be more willing than others to try new efforts. Finally, in the state where outreach to public schools is already an important component of the state university missions, the respondent felt the emphasis on this component just was not a priority (although the outreach requirement exists).

10. *Are their funds available at the state level for teacher training for Ph.D.s?*

- Two – No.
- One – Don't know.

In one state, teacher collaboratives may assist teachers obtain additional education. This state also requires that individuals pay for certification themselves. In another state, some school districts pay for alternate certification and there is a foundation-sponsored pool of funds available to a few candidates.

11. Are there other positions in which you would like to see Ph.D.s contribute?

Number of Mentions	Item Mentioned
2	Curriculum development and curriculum development reviews
2	Teacher training, including preparing new teachers through workshops
2	Use Ph.D.s in expert capacity for focus groups, Delphi panels, expert opinion, etc.

The two interviewees who said they use Ph.D.s in an expert capacity were referring to textbook reviews, summits to inform the Department of Education about priorities, and curriculum development reviews.

Other items mentioned by interviewees included:

- Serve as a resource to teachers.
- Provide training through education service centers in state.
- Professional development schools, where college faculty and school faculty work together in training teachers.

GRADUATE DEANS

Response to Interviews (N = 5)

1. Do you see recent Ph.D.s as a source of quality science and mathematics teachers in high schools [in your state]?

Number of Mentions	Item Mentioned
1	To some extent
4	Yes

2. Why or why not?

Number of Mentions	Item Mentioned
2	Ph.D.s are underemployed and the job market is more challenging
2	Teachers (those with education degrees) don't have disciplinary training and aren't serving the nation well

Other reasons why mentioned:

- Research skills.
- Motivation.

Other reasons why not mentioned:

- Training is not adaptable to teaching.
- Don't have teaching as a career goal.
- Foreign students are not a match.
- Need to re-tool graduate schools and faculty for this.

Other comments:

- Made the link with the community college market now need to expand to K-12

3. What do you see as the advantages of Ph.D.s as secondary school science and mathematics teachers?

Number of Mentions	Item Mentioned
3	Knows subject matter in a deep way (depth of knowledge)
2	Can excite students about science and math
2	They have research and experimentation backgrounds

Other mentions: can serve as role models, Ph.D.s want to teach at some point during their career; K-12 will let Ph.D.s find the right teaching venue for themselves, and can facilitate research in classrooms.

4. What do you see as the disadvantages of Ph.D.s as secondary school science and mathematics teachers?

Number of Mentions	Item Mentioned
3	Pay scale/salary of teachers
2	Credentialing/certification process that Ph.D.s are not tolerant of
2	Secondary school is a challenging environment (e.g., discipline issues)
2	Ph.D.s want to be at universities and don't think beyond academia (nor do their professors support thinking beyond academia)
2	Transition from graduate school (environment and content) to public schools is very difficult

Other mentions included: requires a particular type of personality to fit into the secondary schools; teachers and administrators may be nervous or suspicious of the Ph.D.; unionized school districts; no follow-up from universities, mentoring, or career development; not a prestigious career; Ph.D. has too much knowledge and doesn't know where to start; pedagogy is geared to the adult learner, not youth; and K-12 labs are inadequate at best

5. *What are the obstacles to attracting and retaining science and mathematics Ph.D.s as secondary school teachers?*

Number of Mentions	Item Mentioned
3	Pay scale/salary of teachers (including lack of pay increases over time)
2	Bureaucracy of schools and school systems
2	Difficult/unattractive job
2	It is difficult to enter system (e.g. through certification process)
2	Schools don't support teachers

Other obstacles that received mention included: research does not connect to the classroom; Ph.D.s have not been in a classroom in a long time; the school environment; tremendous time commitment to teaching (multiple periods a day with little down time); professors do not support teaching—they want their students to be just like them; Ph.D.s need pedagogical training for teaching without extending time to degree; few resources and opportunities at the school district level; education (teaching) is at the bottom of the totem pole in terms of prestige; Ph.D.s must be invited in and given a chance to make a difference; need an opportunity to stay in research, be recognized by peers and be in touch with a university; lack of collegiality; school district needs are generally unappealing to Ph. Ds – in rural areas or inner cities; education courses are terrible; and alternative certification is not discussed in universities.

6. *What arrangements have been or could be implemented by graduate schools that would facilitate the transition of science and mathematics Ph.D.s to secondary school teaching?*

Number of Mentions	Item Mentioned
3	Through concurrent degrees or fellowships, offer teaching credential along with science/math degree
2	Fellowships or summer programs that focus on providing graduate students experiences in schools
2	Involve graduate students in providing teacher-focused activities (assistants to teachers, etc.)
2	Fellowships focused on K-12 experiences and possibly teaching credential

Other arrangements mentioned by respondents included:

- Involve graduate students in an existing program that links science faculty with teachers and experienced teachers from high schools.*
- Provide workshops and seminars on alternative careers and including teaching at the secondary school level in these events.*
- Bring back alumni Ph.D.s in science and math who are teaching to present at the alternative careers workshops.
- Provide a loan deferment and loan repayment program when a Ph.D. teaches.
- Shorten the graduate degree program time and decrease time for postdoctorate studies.

- Provide supervised teaching, credentialing, and 2 year postdoctorate as part of a package.
- Universities should transform Schools of Education or bypass them altogether as they are not meeting the needs.
- Establish two tracks—traditional Ph.D. and Ph.D. teaching track; the latter should not end up with a pejorative connotation.
- Have students serve as an apprentice teacher in the classroom; instead of having students complete their TA at the university.
- Urge bachelor's prepared teachers returning for a degree to obtain a content area degree, not one in education.
- Enhance applied science and math departments.*
- Consider a master's degree as an intermediate step (for those who might otherwise stop at a BA).*
- Offer an MA in applied math.*

(Note: Items with the asterisk (*) are those efforts currently implemented or in process at the universities included in the study.)

7. *Are any of these currently available? Which would you be willing to implement?*

   - Yes – four.
   - Probably Not – one.

Rationales for why a university would not be likely to implement certain programs/efforts are degree pressures, faculty time required, funding for degree, American students are in competition with foreign students and teaching will increase the gap, and an overall sense that it "wouldn't work."

In addition to the astericked items in question #6 other thoughts offered by respondents on other programs included:

- The science department at our university led the charge to get the university involved with teaching and education issues.
- Expand public school involvement with existing programs.
- Tried to develop a joint credentialing program with another university but it was not funded.
- Faculty, post-docs, and graduate students all volunteer in schools to work with teachers. Program will expand to student-focused activities.
- Offer an MS in science to existing biology teachers.
- NSF should offer funding to science students to work in K-12.

8. *What incentives or special arrangements have been or could be implemented by school systems to attract science and mathematics Ph.D.s to secondary school teaching?*

Number of Mentions	Item Mentioned
3	Budget for equipment and laboratories

Respondents offered a range of ideas related to incentives or arrangements that schools could implement to attract Ph.D.s to secondary school teaching. These included the following:

- Provide alternative routes to certification.
- Provide summer immersion programs (introduction to teaching, similar to a Berlitz course).
- Provide mentors for teachers in schools.
- Waiver the school curriculum so the specialist can teach in his area of expertise.
- Provide teaching internships.
- School systems should provide opportunities and funding for graduate students to be involved.
- Provide fast track credentialing as part of the graduate program.
- Offer career ladders.
- Emphasize AP and honors courses.
- Provide computer equipment to teachers and ensure they have Internet access.
- Have a better evaluation system.
- Better salaries.
- Provide opportunities for the individual to have an active, scholarly life.
- Provide a budget for student field trips.
- Restructure the school day so there are expanded class times a couple of times a week rather than 5 days of 50 minute periods.
- Concentrate all Ph.D.s in magnet schools (so they are all peers).
- Recruit graduate students from universities for teaching.
- Do not place new teachers into the worst environments in the schools.

9. *Are any of these currently available? Which of these could additionally be implemented by states, school districts, or schools?*

Other ideas offered by respondents (that are currently in existence) include: the Preparing Future Faculty program (national program) that links prominent universities' graduate students with small colleges and community colleges to share facilities and knowledge. The respondents indicated this program could be expanded to high schools. Another university was assisting a school district in developing a health sciences curriculum, funded by the NSF. In one state, a few public and private schools systems had modified the school schedule to allow for longer class periods fewer days a week.

Additional ideas mentioned by respondents included involving the National Endowment for the Humanities in education; examining once again the issue of a master's degree as a “valid” and terminal degree; and experimenting with an on-line AP program where a cadre of Ph.D.s develop internet-based courses and operates the program (using e-mail) -- school districts would pay on a per-student basis. One respondent indicated that it is a self-perpetuating cycle; only the school districts that offer 4 years of science and math generate an interest in students in continuing these studies in college. Because it is not universally available all four years in schools, it becomes the domain of the “nerds” from districts that offered the courses.

DISCUSSION

10. Science and Mathematics Ph.D.s as Teachers

Respondents working at the district or high school level (high school principals, magnet principals, and superintendents) were less likely than state policymakers or graduate school deans to see science and mathematics Ph.D.s as a source of quality secondary school teachers. When queried as to whether science and mathematics Ph.D.s were a source of quality secondary school teachers, respondents answered:

- high school principals: No
- magnet school principals: Maybe
- school superintendents: Maybe
- state policymakers: Yes
- graduate deans: Yes

11. Benefits Ph.D.s May Bring to Secondary School Teaching

- Half of all respondents saw a Ph.D.s content knowledge as a strength.
- Magnet and high school principals saw Ph.D.s as a source for enhancing their schools' credibility and standing, such as with accreditation bodies or in generating needed grant funds.

12. Obstacles to Using Ph.D.s in Secondary School Teaching

- All but one respondent indicated salary was an obstacle to recruiting Ph.D.s to secondary school teaching.
- High school and magnet school principals saw Ph.D.s as overspecialized and overprepared for teaching secondary school students.
- High school principals, magnet school principals, and superintendents were more likely than state policymakers or graduate deans to say Ph.D.s needed educational coursework in order to teach.

13. Graduate School Initiatives

When queried about the types of arrangements graduate schools could implement to facilitate the transition of science and mathematics Ph.D.s to secondary school teaching, respondents typically indicated colleges and universities should offer educational coursework during graduate school and provide opportunities for classroom experience.

- At least one individual within each respondent group suggested Ph.D.s should be provided experience working in classrooms/schools during graduate school.
- At least one individual within all respondent groups (except magnet school principals) indicated Ph.D.s should take educational coursework while pursuing an advanced degree.

- Across respondent groups, interviewees expressed misgivings about whether graduate schools would be willing or able to implement these recommendations—the typical interviewee response was "maybe."

14. Education Courses and Certification

- The respondent category predicted certain opinions. For instance, high school and magnet school principals uniformly indicated Ph.D.s would not be able to teach without educational coursework. Individuals within respondent categories more distant from school operation (state policymakers and graduate deans) were less likely to believe educational coursework was necessary to teaching. When compared to principals and superintendents, state policy- makers and graduate deans were more likely to feel schools of education may not be preparing the highest quality level of teachers needed for science and mathematics.
- Many respondents were not familiar with their state's alternative certification process and how alternative candidates are funded.

15. Other considerations

- Reduced teaching loads for Ph.D.s was seen primarily as a funding issue and one that was not specific to Ph.D.s. A few individuals indicated teacher unions impacted the feasibility of this approach.
- Although only one respondent made them, several comments are worthy of note. The comments are: (1) need role models and people of color to teach science and mathematics at the secondary school level (magnet school principal), (2) there is a need for scientific research at the high school level (magnet school principal), and (3) graduate schools could link with educational professional associations (state policymakers).

Appendix F

Biographies of Committee Members

COMMITTEE ON ATTRACTING SCIENCE AND MATHEMATICS PH.D.S TO SECONDARY SCHOOL TEACHING

COMMITTEE MEMBERS

N. Ronald Morris is a Distinguished Professor in the Department of Pharmacology at the Robert Wood Johnson Medical School, University of Medicine and Dentistry of New Jersey. Dr. Morris is a cell biologist whose lab studies nuclear migration and its regulation, working on the fungus Aspergillus as a model organism. He has previously served as Associate Dean for Research at the Robert Wood Johnson Medical School. He received a M.D. from Yale University School of Medicine in 1959.

Daniel S. Hamermesh is the Edward Everett Hale Professor of Economics at the University of Texas, Austin and research associate at the National Bureau of Economic Research. Previously he has held faculty positions at Princeton and Michigan State Universities, and has held visiting professorships at numerous universities in the Untied States, Europe, Australia, and Asia. He authored *Labor Demand, The Economics of Work and Play,* and a wide variety of articles in labor economics and the leading general and specialized economics journals. Dr. Hamermesh's research concentrates on labor demand, social insurance programs (particularly unemployment insurance) and unusual applications of labor economics (e.g., suicide, sleep and beauty). He currently serves on OSEP's Committee on Methods of Forecasting Demand and Supply of Doctoral Scientists and Engineers. He received a Ph.D. from Yale University in 1969.

Kristina Peterson teaches chemistry and biology at the Lakeside School in Seattle, Washington. She received her Ph.D. in analytical chemistry from the University of Washington in 1997 and served as a distance learning instructor and course designer and as a science advisor to the Teacher Certification Program Wetlands Project at the University of Washington before joining the Lakeside faculty. She has been an active member of the Seattle Chapter of the Association for Women in Science, serving as chair of its program committee, 1995-96 and 1998-1999 and of its scholarship committee, 1994-1995.

Kimberly Tanner is a NSF postdoctoral fellow at the University of California, San Francisco in Science Education (PFSMETE). She is working with the UCSF Science and Health Education Partnership to: 1) determine the factors which contribute to the success of teacher-partnerships; 2) evaluate the impact of these partnerships on participating students, teachers, and scientists; and 3) utilize the results of this research to create materials to facilitate teacher-partnerships that can be disseminated for use by universities and school districts nationwide. Dr. Tanner received her Ph.D. in neurosciences from the University of California, San Francisco in 1997 and received a B.A. in biochemistry/biology from Rice University in 1991.

Mary Long is currently coordinator of UTeach, the secondary science and mathematics teacher preparation program at the University of Texas, Austin. She previously served on the Austin ISD Science Curriculum Team and as the Manager of the District's Science and Health Resource Center, Austin ISD. For a number of years she held positions at the LBJ High School in Austin, beginning as science teacher and advancing to Director of the Science Academy of Austin at LBJ High School. Ms. Long received a M.Ed. degree in science education from the University of Texas, Austin.

Wendy Kopp is the founder and president of Teach For America, the national corps of outstanding and diverse recent college graduates who commit to teaching in public schools. Since its founding in 1989, Teach For America has fielded 5,000 corps members into underserved areas from South Central Los Angeles to the Mississippi Delta. Ms. Kopp holds a bachelor's degree from Princeton University where she participated in the Woodrow Wilson School of Public Affairs. She has received honorary degrees from Connecticut College and Drew University.